THE DEPARTMENT OF

"Michael Belfiore has created a compelling, fascinating, and altogether accessible view inside the past, present, and future of blue-sky defense research. If you want to know who really invented the Internet, or how brain waves can control robotic limbs, or how smart cars will become brilliant, this is your book. A must-read for those interested in invention in the modern age, and those who want to learn how government can foster innovation without creating sluggish bureaucracy."

—Joe Pappalardo, *Popular Mechanic*

"A fascinating introduction to a veritable pantheon of geek gods who quietly shaped the face of modern technology."

—Daniel H. Wilson, roboticist and author of
How to Survive a Robot Uprising

"An entertaining and information-rich account of a small, efficient government agency that often turned twentieth-century sci-fi into twenty-first-century technical reality. Belfiore will inspire young readers of a scientific bent to flood DARPA with their résumés."

—Robert Wallace, author of *SPYCRAFT:*
The Secret History of the CIA's Spytechs from Communism to Al-Qaeda

"An inspiring book about a crucial government agency (DARPA) with a driving spirit to do the impossible and to do it fast. We all need to read this book."

—John Seely Brown, former chief scientist of Xerox
and director of its Palo Alto Research Center (PARC)

Rocketeers:
How a Visionary Band of Business Leaders, Engineers, and Pilots
Is Boldly Privatizing Space

HARPER

NEW YORK · LONDON · TORONTO · SYDNEY

THE DEPARTMENT
OF MAD SCIENTISTS

How DARPA Is Remaking
Our World, from the Internet
to Artificial Limbs

MICHAEL BELFIORE

HARPER

FIRST HARPER PAPERBACK PUBLISHED 2010.

Designed by Suet Yee Chong

The Library of Congress has catalogued the hardcover edition as follows:

Belfiore, Michael P., 1969-
    The department of mad scientists : how DARPA is remaking our world, from the Internet to artificial limbs / Michael Belfiore. — 1st ed.
        p. cm.
    Includes bibliographical references.

    ISBN 978-0-06-157793-2

    1. United States. Defense Advanced Research Projects Agency. 2. Science and state—United States. I. Title.

U394.A75B45 2009
355'.070973—dc22   2009018015

ISBN 978-0-06-200065-1 (pbk.)

11   12   13   14   OV/RRD   10 9 8 7 6 5 4 3 2

For Jade Star

//// CONTENTS ///////////////////////////////

This book could not have been written without the support of DARPA director Tony Tether, DARPA public affairs officers Jan Walker and her staff, and the DARPA program managers and contractors who granted me interviews. I thank them for taking a chance on me and this project. I also want to thank my editor at Smithsonian Books, Elisabeth Dyssegaard, whose keen editorial eye and intuitive understanding of what I wanted to accomplish unerringly guided the ultimate shape of the book. Thanks also to my agent, Linda Lowenthal of the David Black Literary Agency; Seth Fletcher and Bjorn Carey, who assigned me stories on DARPA topics for *Popular Science*; Leander Kahney and Noah Schactman, my editors at *Wired News*; my wife, Wendy Kagan, for her support through another book project; and finally to my good friend Harry LeBlanc for suggesting a variation of the subtitle.

I FIRST MET THE AUTODOC in a science fiction novel. Back then it was on a starship, and it allowed its user to live practically forever—pretty much a requirement for interstellar travel, even at near-light speeds. Now I was seeing it in the flesh, as it were, squatting in a corner of a Silicon Valley laboratory, looking like a giant mechanized insect with its mouth parts and legs poised over an operating table.

The body on the table, like the robot, was artificial—placed there to show what an actual human being would look like in that setting. My tour guide was a forty-seven-year-old mechanical engineer talking a mile a minute, pointing out a cabinet full of surgical supplies and a swing-arm nurse built on the frame of an industrial robot used to working more on cars than on human bodies, and explaining how the system worked. It didn't take much imagination to bring the scene alive in my mind's eye.

The patient, gushing blood, near death, would be tossed onto the table, none too gently, given the circumstances: massive hemorrhaging that had to be stopped immediately. Strapped down, breathing shallowly, the injured man wouldn't flinch as the robot maneuvered its grasping parts over his wounds. The mini CT scanner attached to the table would slide down the length of the man's body, feeding data to the robot's computer brain. Then the robot would get to work, first jabbing the patient with a needle, then deftly threading a line through his vascular system, guiding it unerringly to the weakly pulsing artery that was pumping the man's blood out over the table and on to the floor.

A pause, while the robot worked invisibly inside the man's body. And then the flow of blood would shut off as abruptly as a turned faucet as the robot plugged the leak. Whirring and clicking, the robot—apparently satisfied—withdrew its graspers and pulled out the bloody line, replacing it with a simple IV. At an unspoken command, the robot nurse swung over to the supply cabinet, picked up some sutures, and swung back to the robot surgeon. Then it pulled one of the surgeon's arms off and plugged in a fresh one. The surgeon picked up a suture with its new hand and made quick work of sewing the patient up. Time elapsed from diagnosis to completed surgery: approximately two minutes. And another soldier would live to fight another day.

"Want to try it?" the engineer was asking.

"Of course," I said.

He seated me at a table before another robosurgeon—the latest prototype—helped me position my hands on a pair of metal grips, and had me lean forward so that I could look through the binoculars on the console in front of me. I was instantly transported across the room to another table, where a slab of simulated flesh awaited my ministrations with needle and suture of my own. I saw the table and its simulated patient in full 3-D color, and my hands registered every bump and jerk as I less than expertly maneuvered my robot's graspers and tried to sew. I wasn't just watching a science fiction novel come to life now, I was *in* one.

My visit to the robot surgeon was just one stop on my odyssey through the startling and mind-bending world of an obscure government agency almost no one I speak to casually has heard of but that has affected all of our lives in countless ways. Have you used the Internet? A computer mouse? A satellite-based navigation system? Thank the Defense Advanced Research Projects Agency, or DARPA. My mission in writing this book was to find out what projects the agency was working on today that could prove as influential as those past projects, and meet the people bringing them to life.

. . .

I FIRST STUMBLED on DARPA while reporting on the world's first privately funded manned space flights. It was June 22, 2004, in Mojave, California. The TV crews had cranked down their satellite dishes and folded their tents and driven back to L.A. in their trucks and vans, followed closely by the thousands of spectators in their RVs and cars. The windswept little town in the Mojave Desert one hundred miles east of Los Angeles had, in just a day, returned to its usual near-emptiness, punctuated by the sound of the wind whipping flagpoles and the occasional roar of jet engines and the growl of piston engines from airplanes taking off and landing at the airport. Which was now a spaceport.

The day before, *SpaceShipOne* had carried a man, sixty-three-year-old Mike Melvill, outside the atmosphere for the first time. The ship had been conceived by an airplane designer renowned for his shockingly in-novative approach to engineering and built by his small company, Scaled Composites, headquartered on the airport's flight line, for a mere $25 million. That was half the retail price of a Boeing 737-600 airliner. The little spaceship that could, hand-built of carbon-fiber composites, flew three times the speed of sound, faster than any civilian craft yet built. While still far slower than the Mach 25 speed necessary for reaching orbit, *SpaceShipOne* nevertheless breached the atmosphere and allowed Melvill to float, weightless, for four minutes. News of the achievement now blared in headlines around the world. But I alone, it seemed, among the journalists covering Melvill's flight, had returned to Mojave the next day to see what a more or less ordinary day looked like at a commercial spaceport. There was no one around. The place might as well have been a ghost town. Everyone was inside, out of the desert sun, working in their hangars and offices during just another day.

I began by checking out a rocket company called XCOR Aerospace. As it turned out, that was as far as I got that day. Though one would never guess it from its outward appearance, XCOR Aerospace represented the hopes, dreams, and potential of an age of private space travel as well as any other group of aerospace mavericks, including Scaled Composites. XCOR Aerospace occupied a featureless blue hangar dating from the Second World War. A sign on the door featuring a stylized rocket plane logo was all that distinguished the place. I walked into a battered lounge that was trying without much success to be a lobby. Framed magazine article reprints about the company hung on the walls, including pieces about a home-built airplane the company's engineers had hot-rodded up with a pair of rocket engines and dubbed the EZ-Rocket.

The heart of the operation was an open hangar in which working areas and workbenches had been laid out along the sides. There I met a friendly middle-aged engineer named Aleta Jackson, who I later learned possessed a keen intellect and not a little of the sense of humor required of anyone who has devoted her life to the fulfillment of a dream most people think impossible. She was about as far removed as one could imagine from the stereotypical picture of the fabled steely-eyed missile man of the previous generation, who had started the first space age, the one run by massive government programs, but she was no less committed to the dream of space flight. "Without routine, regular, reliable access to space, I think this planet is hosed," she told me simply. "We've got to have it, and the sooner we get it, the better off everyone's going to be."

Everyone?

"You can do so much more if you have routine access to space," she explained. "You can bring back resources that we don't have."

For instance?

"One carbonaceous chondrite asteroid three kilometers in diameter will probably produce as a by-product more gold than has been mined to date on the Earth."

XCOR's chief engineer, Dan DeLong, likewise subscribed to an all-encompassing belief in the transformative power of affordable access to space. "It's almost a religious thing," he told me when he joined me and Jackson there in the XCOR shop. "You have to believe. You have to be an optimist. Obviously if we wanted to start a company just to make money, we would have done something much easier."

Lofty though their dreams might be, the rocketeers of XCOR took a very practical and workaday approach to the problems of affordable access to space. They needed investors, so they built their EZ-Rocket technology demonstrator, showing not only that their dreams were built on sound engineering principles, but that rockets could be safely flown as an ordinary part of a viable business. Further demonstrating the safety and reliability of their rockets, the team built a "tea cart" rocket, a small, fifteen-pound thrust, ethane-and-nitrous-oxide-fueled rocket on a rolling cart similar to those used to wheel room-service meals to hotel rooms. In fact, the team actually did fire their device in a hotel ballroom. "And we did it with fire marshal approval," DeLong said, rightly proud of the achievement. "The fact that we had fire marshal approval is what impressed the investors."

I found this juxtaposition of outlandishly ambitious goals with basic, commonsense business principles and everyday problem-solving inspiring. Unlike Scaled Composites, XCOR did not have a single angel investor with plenty of disposable cash. Its investors expected a return on their investment, it was hoped sooner rather than later. That didn't bother Jackson, DeLong, or their colleagues. Fresh from the disappointment of being part of a rocket company called Rotary Rocket that had burned through some $30 million in financing without a working spaceship to show for it, the group now subscribed to a "small is beautiful" philosophy. "Having too much money probably causes more companies to fail than having too little money," DeLong asserted. "You never learn thrift if you start out with too much money."

The XCOR managers' solution to their cash-flow problem was to find individual customers who also needed some of the technologies XCOR wanted to develop. Their relatively modest first space milestone was a two-seat rocket plane that would take off from a runway on rocket power, without the encumbrance of a jet-propelled mother ship to reach the same kind of altitudes and speeds as the jet-plane-plus-rocket-plane *SpaceShipOne* system. The rocket plane, Jackson and the others hoped, would find markets among space tourists and among scientists wishing to fly microgravity experiments more cheaply than had ever before been possible. Until they got there, the XCOR engineers hoped to turn enough of a profit in developing some of the technology behind their planned spaceship to fund the development of components for which they either could not or preferred not to find customers.

At the time of my visit, the XCOR engineers were well along this path toward commercial spaceflight. They were then finishing development of a piston-driven liquid oxygen pump. Not exactly the headline-grabbing project that *SpaceShipOne*'s flight was, but in its way, an important development nevertheless, at least as far as people who understood rockets were concerned. People such as the program managers of an obscure research-and-development agency called DARPA. DARPA had invested $750,000 toward getting XCOR's new pump to work, and now they were watching their investment pay off.

At the time I didn't register much about DARPA other than that it was one of XCOR's major customers and that it was a government agency. I had started writing a book about small entrepreneurial spaceship companies such as Scaled Composites and XCOR, and it seemed to me that the private sector was where the action in space was, not government programs.

The commercial space flight industry had gotten off the ground with the $10 million Ansari X PRIZE for the first privately funded space shot. The prize stipulated that the winner receive no funding of any kind from any government; the prize's founder, Peter Diamandis, believed that the main reason space flight wasn't more affordable and commonplace was that government space programs had stifled the innovations that would have made it so. The X PRIZE competitors were inclined to agree with him, and writing about the X PRIZE competition and its aftermath, so was I.

And yet, much as I sought to dismiss it, DARPA kept popping up as

I researched my book *Rocketeers: How a Visionary Band of Business Leaders, Engineers, and Pilots Is Boldly Privatizing Space*. There it was funding Space Exploration Technologies Corporation, or SpaceX, founded and led by Elon Musk near the Los Angeles International Airport. Musk had made a fortune by co-founding PayPal and then selling it to eBay in 2002, and now saw space travel as humankind's best hope for surviving the kind of environmental upheavals that had wiped out the dinosaurs. Musk wasn't just shooting the moon—he wanted to go all the way to Mars, and hoped to provide the technological infrastructure to allow others to settle there permanently. No mere pipe dream, SpaceX had $100 million in working capital contributed by its founder alone, and it was far along in building and testing a rocket called the Falcon 1. Musk's team of highly competent engineers, many poached from SpaceX's big competitors, also believed the plan would work. And so did DARPA, which became SpaceX's first customer, for a demonstration launch of Falcon 1.

And DARPA was funding yet another space start-up, AirLaunch LLC, founded by former Rotary Rocket CEO Gary Hudson, whose mission was to drive down the cost of launching satellites by dropping rockets off the back of C-17 cargo planes similar to the way *SpaceShipOne* dispensed with the expensive throwaway first-stage boosters of its competitors by using an airplane as the first stage. For a crew launch, instead of kicking this larger rocket out the back of a C-17, AirLaunch would sling it from the belly of an airplane custom-built by Scaled Composites, or, in a pinch, from a specially modified Boeing 747. Through a spin-off company called Transformational Space Corporation, or t/Space, Hudson tried to get NASA to fund his and his partners' manned space-flight scheme. After paying t/Space $6 million to study the idea, NASA let it drop. Hudson wasn't especially surprised by NASA's refusal to follow through on an entirely new idea for launching people into orbit. His feeling, like that of many of the entrepreneurs working in what was coming to be called the NewSpace industry, was that NASA was too bureaucratic, too stuck in a cold war–era mode of operation favoring the big-ticket contractors to consider a fresh idea just because it could be less expensive (and maybe even safer, to boot). Hudson merely shook his head and went back to work at AirLaunch, happily test-firing rocket engines at the Mojave Air and Space Port and dropping dummy rockets out of C-17s. AirLaunch, as I found out, was an Air Force program managed by none other than DARPA.

How had DARPA established such close working relationships with some of the most conceptually "out there" among the space entrepreneurs, who were nevertheless very well grounded in their approaches to problems that conventional "wisdom" said were insurmountable on anything less than the budgets of major government programs? This had to involve two major leaps on DARPA's part, it seemed to me: (1) the flexibility to see that difficult problems might be solved by not just unconventional means but also unconventional people, and (2) the insight to sort out the truly viable approaches from the mere crackpot schemes. What exactly was this DARPA, I wondered, and why did it appear to want cheap access to space as badly as these desert renegades?

Eventually I circled back to give DARPA another look. What I learned astonished me. DARPA, I discovered, was in fact America's first space agency. That's right. Not the National Aeronautics and Space Administration, America's civilian space agency, but DARPA—conceived as the research-and-development arm of the Department of Defense—initially took up the call to arms issued by the administration of President Eisenhower to meet, and then surpass, the Russians in the high ground of space.

Eisenhower established both NASA and DARPA—then known as ARPA, without the *D* for *Defense*—in response to the Soviet Union's 1957 launch of *Sputnik*, the world's first artificial satellite. But while NASA's charter focused squarely on beating the Russians in space, ARPA's mission gave it a broader purview, taking in all advanced technologies in its mission to guard the United States against technological surprise from any quarter. ARPA gave some of the early U.S. rocket programs a significant boost before NASA, accreting mass like a young star forming a solar system around itself, absorbed those programs, and eventually beat the Russians to the moon. Meanwhile, ARPA quietly moved into areas such as information technology. In 1969, the year that NASA landed the first people on the moon, ARPA launched the little-noticed ARPANET—which arguably was to have an even greater impact as the backbone of the Internet.

In the 2000s, even as NASA continued to search for a clear purpose more than thirty years after achieving its original mission, DARPA got back into space in a major way. Its program managers wondered what possibilities would present themselves if only access to space could be more "operationally responsive," in military parlance. How about pop-up satellites, launched on specific spy and communications missions in response to particular conflicts? Could the air force or army send special forces on a strike mission on the other side of the world with just a couple of hours' notice?

The more I dug, the more I found myself fascinated by this deliberately low-profile, but hugely influential, Department of Defense arm. Here was a government agency—with a $3 billion budget, no less—that acted like a small company, that got innovative technology projects funded quickly and with a minimum of bureaucratic hassle. How did it pull off this seemingly impossible feat? What lessons lay there for other companies and organizations? And, even more enticingly, what quiet, unheralded DARPA programs in progress might prove as influential as the Internet?

There was yet another reason I was intrigued by DARPA. The United States spent $651 billion on defense-related activities in 2009. That's more than half of the U.S. discretionary budget, and more than the military budgets of China, Russia, and Europe combined. No less a figure than President Eisenhower—the same president who launched NASA and DARPA—warned of the direct cost to the nation's citizens of extreme military spending. "Every gun that is made, every warship launched, every rocket fired signifies, in the final sense, a theft from those who hunger and are not fed, those who are cold and are not clothed," he told the American Society of Newspaper Editors in a speech that was broadcast over television and radio in April 1953. "The cost of one modern heavy bomber is this: a modern brick school in more than 30 cities. It is two electric power plants, each serving a town of 60,000 population. It is two fine, fully equipped hospitals. It is some 50 miles of concrete highway. We pay for a single fighter with a half million bushels of wheat. We pay for a single destroyer with new homes that could have housed more than 8,000 people. . . . This is not a way of life at all, in any true sense. Under the cloud of threatening war, it is humanity hanging from a cross of iron."

Harsh words for what the president was later famously to call the "military-industrial complex" in his 1961 Farewell Address, and all the more startling coming from one who knew better than most the value of a strong, well-equipped, and prepared military. Eisenhower had, after all, commanded the Allied forces in Europe during World War II.

Eisenhower himself saw no alternative to the vast arms buildup and its threat to the health of the nation. "Until the latest of our world conflicts," he explained in his Farewell Address, "the United States had no armaments industry. American makers of plowshares could, with time and as required, make swords as well. But now we can no longer risk emergency improvisation of national defense; we have been compelled to create a permanent armaments industry of vast proportions."

But he did see a way to mitigate some of the ill effects of massive defense spending. In 1956, Eisenhower signed into law the National Interstate and Defense Highways Act, setting into motion the building of the nation's highway system, which came to be known as the Eisenhower Interstate System. Ostensibly to be used for the rapid movement of troops and the evacuation of civilians from cities in the event of an attack on the United States, the Interstate System in fact became perhaps the single most generally useful project of Eisenhower's two terms in office. It was also a way to release a portion of Pentagon funding for the public as well as military good. ARPA has similarly turned out to be a silver lining in Eisenhower's military-industrial complex.

From the creation of the ARPANET, which expanded to become the Internet; to the Global Positioning System, whose precursor system began by showing the way for warships, airplanes, and ground vehicles, and that now guides untold numbers of hikers, emergency workers, and cell phone users; and in countless projects in a breathtaking range of fields, DARPA has fostered and brought into existence some of the most useful technologies of the last fifty years. At the same time, operating on an annual budget only a tiny fraction of the overall defense budget— about the price of one and a half B-2 bombers, or only about one-sixth NASA's annual budget—DARPA today proves that the U.S. military *can* maintain its edge without breaking the back of the economy that tries to support it.

Since a single book on this subject can only scratch the surface, I've narrowed my focus here to a few of the agency's current projects that have the potential to influence our society as much as the Internet and GPS, while at the same time illuminating some of DARPA's extraordinary history. Join me, then, for a wild ride through the back roads of extreme science and technology. DARPA's story is in the best tradition of mad science. Mad because only a few mad dreamers dare to believe in the impossible. Mad because only truly original ideas are judged by others to be crazy until they are proven otherwise.

AN ARM AND A LEG

THE TWIN-ENGINE RIVERINE craft sliced through the Euphrates River a little more than one hundred miles as the crow flies northwest of Baghdad. Marines attached to the First Battalion, Twenty-third Marine Regiment, on board scanned the shoreline for signs of trouble. There, in a palm grove where the river made its first of many bends on its meandering way from the Haditha Dam to the Persian Gulf, another patrol had been attacked by insurgents just minutes earlier, and these marines aimed to weed them out. The marines were based at the dam itself. Their job was to protect the facility that generated fully a third of Iraq's power and to keep the stretch of river above and below the dam clear of insurgents. It was New Year's Day 2005.

The pilot beached the craft on the sandy bank of the river, and the marines jumped out, rifles at the ready, fanning out as they headed into the grove. The battalion's engineer, thirty-three-year-old reservist Captain Jon Kuniholm kept a wary eye out for improvised explosive devices, or IEDs. The insidious roadside bombs had been taking a heavy toll on U.S. soldiers in Iraq since the war began in 2003, and Kuniholm, with the help of his design start-up in his hometown of Durham, North Carolina, had been building a robot that could move ahead of patrols and defuse bombs before they went off. Finding the carefully hidden bombs was an altogether different task, however. They could be concealed in anything, buried anywhere, triggered by anything from a cell phone to a garage

door opener—like the one in the discarded olive oil can that Kuniholm
had just enough time to register before it exploded.

The blast blew him off his feet. He lay dazed as insurgents opened
up with automatic rifles, machine guns, and rocket-propelled grenades.
Kuniholm struggled to rise, searching for his M4 carbine, which, it
turned out, had been torn in half by the explosion. And that's when he
saw that his right forearm was dangling from the rest of his arm below
the elbow on a strip of flesh only a couple of inches wide. "Fuck," he said.
He got to his feet and, holding his severed right hand in his left, he ran
to the cover of a nearby agricultural pump house, where he waited for
the aid of a corpsman.

By the time the rest of the marines were able to beat an orderly retreat
back to the patrol boat, one of their number had been fatally injured and
several others had received lesser wounds. The day had not gone at all
well for the patrol, and with the adrenaline rush that had been keeping
him going wearing off, and grayness creeping in around the edges of his
vision, Kuniholm told the others that he wasn't feeling so hot, and that if
everyone was on board, they should get back to their base at the dam.

So began Kuniholm's long road home, from the base at Haditha to
a field hospital at Al Asad, where surgeons finished the job the IED had
started by taking off the rest of his forearm, then to Germany for more
treatment, and finally home to North Carolina, for yet more surgery
at Duke University Medical Center. From there it was on to the army's
Walter Reed Medical Center, in Washington, D.C., to be fitted with a
prosthetic arm. Actually, a trio of arms.

Like veterans since World War I, Kuniholm got a simple body-oper-
ated hook that he could open and close with the shrug of a shoulder or
by extending the remaining portion of his arm. His movements pulled
on a cable that was attached to the hook at one end and to the harness
that held the prosthetic in place at the other. Great for everyday use, and
durable, the thing lacked a certain panache, however—which Kuniholm's
so-called myoelectric arm sought to provide. Heavy, relatively fragile,
and limited in function, the myoelectric arm nevertheless represented

the state-of-the-art in prosthetics. Electrodes embedded in a flexible liner worn beneath the arm's carbon-fiber sleeve picked up the electrical signals generated by the firing of the residual muscles in Kuniholm's forearm. Microchips translated those signals into commands to open and close the fingers of the prosthetic hand, which could take the form of either a hook or a cosmetically appealing but less functional hand. The device was more lifelike in appearance than the hook prosthesis, and was therefore less intimidating to able-bodied onlookers. But it couldn't be worn long without discomfort, it had to be kept from getting wet and dirty, and Kuniholm generally got less use out of it than the hook or the third arm, which was shorter than the other two and suitable for holding a pencil or pen. His experience with the myoelectric arm was typical. Michael Weisskopf, a journalist who lost his right arm below the elbow while riding with troops in Iraq in 2003, wrote of his struggles with his own myoeletric prosthetic arm in his book *Blood Brothers*.

> *If my former right hand floated lightly the fake one moved like a dumbbell—fat, clunky, and heavy. Its two and a half pounds were concentrated in the electronic hand—the place farthest from the half forearm. The prosthesis made my arm crook out like Popeye's; my range of motion was so limited that I couldn't raise the hand within a foot of my mouth. I kept bumping it into things. I gave up on long-sleeved shirts. They didn't fit over the bulging battery box or couldn't be buttoned over the thick prosthetic wrist. I named it Ralph, after the clumsiest kid in my grade school.*

Back in his design shop, Kuniholm and his colleagues took his prosthetic arms apart and were less than impressed by what they found. They figured they could do better with a more functional design, and they made prosthetic design a major part of the work of their company, Tackle Design. Inspired by the open source (i.e., free) model of software development, they launched an online design forum called the Open

Prosthetics Project, inviting contributions from anyone who cared to make them. And Kuniholm joined a DARPA project called Revolutionizing Prosthetics.

Revolutionizing Prosthetics was started in 2005 by DARPA program manager Geoffrey Ling. An army colonel and intensive-care unit doctor, Ling is one of just 10 percent of DARPA program managers who also serve as active-duty military officers. He shrugs off the suggestion that this makes him unusual at an already extraordinary place to work. "All of us have a research background," he told me of himself and his military colleagues at DARPA, "so I don't think we're all that different from our civilian counterparts in terms of our background and training. It's just that we happen to wear a uniform."

But that uniform has led him to places that many of his civilian colleagues must learn about from afar as they seek ways to better equip their uniformed "customers" (for lack of a better word). Ling has served two tours of duty in combat zones, one in Afghanistan, the other in Iraq. He served the latter tour after he joined DARPA, which makes him even rarer—an active-duty DARPA program manager who has served in both capacities in wartime in a combat zone.

All of which far more than informs his work at DARPA. It utterly defines it.

Colonel Ling is a Chinese American born in Baltimore and raised in New York City. He bursts with energy when he talks of his life's mission to care for wounded soldiers. The words tumble out as fast as he can form them. Time, you get the feeling from listening to Ling, is most definitely not on his side, and like so many others at DARPA, he spends his life in fast-forward. More than that, time's not on the side of the wounded he cares for—young men and women in uniform (and in many cases, even children), wounded in combat. His signature program, Revolutionizing Prosthetics, he likes to say, is not a science project; it's something we have to complete right now, to help these people.

Ling is unapologetically patriotic. "And if I sound like a flag-waver," he tells me, "tough. That's what I am. I mean, I'm not a sixties guy who wants to spit on the flag and spit on soldiers when they come back. That's a lot of bullshit." His deepest passion is to help the young men and women giving their all for their country. The politicians at home might put the cost of the ongoing wars in Iraq and Afghanistan in terms of the tremendous drain on the U.S. Treasury, but for an appalling number of those who returned from their deployments, the war quite literally cost an arm or a leg. Each one of those soldiers, Ling feels strongly, has to—*simply has to*—receive the very highest level of support. "We have got to put our resources and take care of [the soldier]. Because he's representing us in the most positive way possible. Forget the politicians. It's young Americans like that who represent us." Ling's program is dedicated to the soldiers, but he also feels strongly that the technology has to be shared with the rest of the world. "And we're going to do it," he says. "It's America again doing the best things that America can do, which is showing that . . . we are a superpower that really tries to take care of the world."

Ling graduated from Cornell University, in Ithaca, New York, with a doctorate in pharmacology, in 1982, and then went on to earn his medical degree from Georgetown University, in Washington, D.C., in 1989. Then he joined the army, "which was really kind of great," he says. "The army was very good to me. I'm one of those happy guys." After he finished basic training, he got assigned to the Uniformed Services University of the Health Sciences, the military medical school in Bethesda, Maryland. There he treated patients, taught med students, and ran a research lab. He completed a neurology residency at Walter Reed Medical Center and then trained in neurocritical care at Johns Hopkins University, developing a specialty in caring for traumatic brain injury in wounded soldiers.

In late 2002 he got "that mysterious call to come join the club," as he describes the typical way in which program managers get recruited to DARPA. In this case, it came from navy commander Kurt Henry, also an intensive-care unit doctor. "Geoff," he said, as Ling recalls it, "you know, this is actually a pretty good tour of duty. You ought to try it." Ling had heard of DARPA as a place that fostered groundbreaking scientific research, and he said he'd consider it. That was all the encouragement Defense Sciences Office director Michael Goldblatt needed to begin actively recruiting Ling. "That's when the whole process of interviewing began," Ling told me later.

Ling was still on the fence about joining DARPA when he met some-one who completely altered the course of his life. In the fall of 2003, the army sent him to Afghanistan, where he joined the Forty-fourth Airborne Medical Brigade in Basram. There, most of his patients were civilians. "I took care of a lot of children," he told me later. "And those children were blown up primarily by old Russian land mines." The kids would find the things on the ground, begin playing with them, and then they'd explode, blowing off fingers and arms and legs.

One day a patient came in, a little boy, "just the cutest little boy that there ever was," Ling told me. He had lost a leg and most of his right hand to a land mine. Ling would never forget that boy. It struck him that be-cause the kid was otherwise unharmed, he had a long life ahead of him. But what kind of life would it be? "Without the use of one arm and one leg," Ling told me, "in this kind of environment, you know his life would just be tragic." That got him thinking, hard, not only about the children he was seeing every day in his field hospital but also about a lot of the American soldiers under his care. These were people who had lost limbs but who would otherwise make a full recovery.

That little boy stayed in his mind after his return from Afghani-stan, and he knew that he had to join DARPA, had to do something for people like him, had to help them to live normal lives. DARPA, he believed, was the one place where he could help bring about the kinds of advances in medical technology that could allow patients like these to return to something approaching their normal lives after their wounds had healed. "DARPA had the money," he explained to me later. "DARPA had the minimal bureaucracy, DARPA had the right attitude, DARPA had the stated mission of pushing back the frontiers of science. I came back really invigorated to do this. I really, really, really wanted to come to DARPA." Fortunately, DARPA director Tony Tether signed Ling up in June 2004.

Once again, however, fate intervened, this time in the form of a deployment to Baghdad in 2005. It was an eye-opening experience. Ling was with the same unit as in Afghanistan, serving with the Eighty-sixth and Tenth combat support hospitals. As in Afghanistan, some 80 percent of the patients he cared for were not Americans. This time, most of them were Iraqi military police and soldiers, who at that time comprised most of the casualties.

One of his American patients, a twenty-two- or twenty-three-year-old Humvee gunner, arrived at the hospital in a helicopter with his back broken in three places. His Humvee had hit an IED, and he'd been thrown from the vehicle. Fortunately, the man's spinal cord was still intact and he could move his legs and his toes. So Ling told him, hoping to cheer him up, "Specialist, you're going to go home now. You've got a million-dollar wound."

To Ling's astonishment, the soldier began to weep.

"You don't have to feel bad about that," Ling told him. "You don't. Because you got injured doing very honorable work and you're one of America's young heroes."

The young man was not mollified. He grabbed Ling's jacket and said, "Sir, that's not why I'm upset." Back home, he told Ling, he felt he had no real purpose. He was the manager of a fast food restaurant. But in Iraq, he believed he was doing important work, helping people rebuild their country.

Ling himself was almost moved to tears by the soldier's passion. The episode stayed with him every bit as much as his experience with the wounded little boy he'd treated in Afghanistan. Because, he explained to me later, it gave him a good hard look at why soldiers were there in Iraq. "Forget the reason why anybody else thinks they're there," Ling told me. "... *They* think they're there ... for the reasons that America is such a wonderful country—that we will go out and help other people build their own country. Not take it over. Not take all the oil for ourselves.

Not subjugate their people. But really to go in there and help build their own country."

Jon Kuniholm, the marine who lost his arm in Haditha, might beg to differ with that last statement. After his wounding, he came out strongly against the war in Iraq. "Marines do not pick and choose missions," he said in a speech at the Democratic National Convention in the summer of 2008, "our leaders do. I believe the war was a bad idea and poorly planned at the highest levels." He, like most of his comrades, would serve his country again with pride if called upon to do so, he said, "but our sense of duty and sacrifice do not validate our leaders' decisions. The notion that criticism of the war dishonors our sacrifice makes no sense."

Still, Kuniholm was in total agreement with Ling on at least one essential point—that those who served their country should in return receive their country's service. Ling's tours in Afghanistan and Iraq, he said, "absolutely validated my life, and told me that I have to go back to DARPA and do all the right things for kids like this."

The result was DARPA's Revolutionizing Prosthetics program. Ling's goal for the program was nothing less than the creation of an artificial human arm that looked, behaved, weighed the same as, and could be controlled by its user like a native arm. Ling's focus on an arm rather than a leg was an obvious choice for a program that sought to revolutionize the technology of artificial limbs. The problem of artificial legs had largely been solved, as far as Ling was concerned. "If you lose your leg, we can replace your leg with an artificial prosthetic limb that will give you such a high level of function that you can do almost everything." Take the case of South African sprinter Oscar Pistorius, who was initially barred from competing with able-bodied sprinters in the 2008 Olympic Games, not because he was handicapped, but because Olympics officials actually deemed his carbon-fiber legs an unfair advantage over his merely fully flesh-and-blood competitors.

Ling isn't one to put down the advances made in this area. "But, really," he says, "what are you asking a prosthetic leg to do? Stand and move, correct? You're not asking to do more than that, and to do that, you only need a hip, a knee and an ankle, and a big toe. That's all you need. And so, from a functional standpoint, it's actually a very straightforward problem to address. Not easy, but a straightforward one."

Prosthetists have tackled the challenge of building prosthetic legs in a number of ways, depending on the task required of them. For a competitive runner such as Pistorius, designers have settled on a curving spike of carbon fiber that acts pretty much like a leg-size spring, helping to propel the runner forward as he swings his remaining thigh muscles back and forth. For day-to-day activities, amputees such as Pistorius have a choice of much more complex machines, essentially leg-shaped robots designed to sense and compensate for the motions of the thigh, depending on how they use them (for instance, to walk up stairs or just stride along on level ground) by rotating and flexing their robotic ankles and knees accordingly.

A similarly useful prosthetic arm is an entirely different engineering prospect. Whereas our legs are ordinarily not much more than our means of getting around, our hands are nothing less than our most versatile tools. Their opposable thumbs built our civilization. They helped paint the Sistine Chapel and sculpt Michelangelo's *David*. They are called on by turns to play Beethoven's symphonies and to haul rocks around a construction site. The problem of replacing an entire arm, from shoulder to fingertips, is especially challenging, since that uniquely, amazingly useful tool that is the hand has to do its work at the farthest end of the arm from its attachment point on the body. Prosthetic legs can simply take their cues from the motions of the user's remaining thigh muscles and from other body motions. But there are far too many muscles—twenty-seven in all—controlling the motions of a hand for a prosthetic to guess at the user's intentions based on the motions of his or her remaining muscles. The remaining muscles in the case of a full-arm amputee are just the chest and back muscles. "That's why prosthetics can't give you much more than a nominally functional hook, or something that looks like a hand but is totally not functional," says Ling. That's all they can do because they just don't have enough residual muscles . . . to control this very complex hand that you and I have."

Actually, there is another reason prosthetics haven't progressed much beyond the simple hooks in use since World War I, which Ling left unsaid, but that is implicit in the very fact that major advances in prosthetic arms have to be funded by the government rather than private enterprise. The reason is simple economics.

Of the estimated 1.7 million Americans who live with limb amputation, only 100,000 or so, about 6 percent, have lost all or part of an arm. Of those, most are older, less active people who have lost their limbs to illness, typically diabetes and the poor circulation caused by that disease. These are people whose rock-hauling, Beethoven-playing days are largely behind them. Not like the veterans returning from Iraq and Afghanistan missing limbs and with most of their lives still ahead of them. But veterans with limb loss numbered only about 700 at the end of 2007, and only about 150 of those had lost all or part of an arm. All of that added up to a relatively tiny market for prosthetic limbs in general, and arms in particular. "When I was going around to different companies asking if they'd be willing to help out, they all asked the same question," says one prosthetic arm specialist looking for support for innovations that could help her patients. "How many will I sell in a year? If there's not a lot of potential profit, they're just not that interested."

Not that people such as Kuniholm blame the manufacturers. Kuniholm is philosophical about the lack of commercial support for prosthetic arms. "Prosthetics is one of many underserved markets in which innovation has stagnated because the traditional incentives are lacking," he has said. "The people who make innovations in this field are usually passionate users tinkering around in their garage."

Hence the need for government intervention. Barring an infusion of capital from a wealthy investor with a sudden need for a decent prosthetic arm, there was simply no other way that major advancements in prosthetic arms are going to be funded. The military's interest in better prosthetics increased with the wars in Iraq and Afghanistan in 2003.

Advances in body armor and trauma medicine enable soldiers to survive wounds that would have killed them in previous wars, but at a terrible price: their arms and legs take the brunt of bomb blasts and weapons fire. Responding to criticism of its treatment of veterans returning from the Gulf War with amputations in the early 1990s, Walter Reed began issuing the myoelectric arms as well as computer-controlled prosthetic legs called C-Legs. The legs served the vets well; the arms, far less so. The will was there to provide returning vets with whatever equipment they needed to return to as close to their normal lives as possible. But the right equipment simply didn't exist; it had to be developed.

One of the major problems with a conventional myoelectric arm is that it depends on electrodes pressed against a user's skin to read the electrical signals generated by underlying muscles. A user must flex his or her residual arm muscles in just the right way to open and close the hand or hook or to cause it to rotate. The socket containing the electrodes has to fit extremely tightly against the user's stump, which, besides just being uncomfortable, tends to restrict the user's movements. Even so, accidental flexing and misread signals often cause the prosthesis to behave erratically, with consequences ranging from simply embarrassing, as when author and journalist Michael Weisskopf's hand rotated through a 360-degree turn in a social setting, to downright dangerous, as when his hand clenched the wheel of his car in a death grip and refused to let go as he tried to make a turn in rush-hour traffic.

Weisskopf eventually settled on a hook attachment for his myoelectric arm for everyday use. More functional, lighter, and thus more comfortable than the cosmetically more appealing hand attachment, the hook became, he said in his book *Blood Brothers*, his "trademark," even a source of pride for him, because of its bald expression of his disability, stares and sidelong glances from strangers be damned, or even welcomed.

Kuniholm dispensed with myoelectric prostheses for everyday use entirely. "I prefer the hook," Kuniholm told me, "because . . . I can open and close the hook faster than I can a myo-hand. And it's a little bit more reliable. You know, it opens for the most part when I want to open it, and it closes when I want to close it." Like Weisskopf, Kuniholm bemoaned the loss of range of movement imposed by the myoelectric arm. The only real advantage he could see to his myoelectric arm was that he didn't require the cable-and-harness system of the body-powered hook. With the myoelectric arm, he told me, "I don't have what amounts to a piece of clothing that I have to wear every day and wash in the sink." The appeal of a more cosmetically pleasing hand wore thin pretty quickly, though. "There have been a lot of sacrifices made in the name of cosmesis and appearance to function," he told me of the myoelectric hand, chief among them, the myoelectric hand's posture was frozen in a slightly cupped position, known as a palmar grasp, which looks nice in public but doesn't provide nearly the utility of a good old-fashioned hook. In some ways, the hooks were actually more versatile than their myoelectric counterparts because they were unburdened by the need to appear natural and could be designed purely for strength and utility, rather than having to incorporate such encumbrances as fake skin and five fingers—only a few of which could actually be used.

Kuniholm figures that fully half of his fellow vets returning from Iraq and Afghanistan with arm amputations prefer their body-powered hooks to their ostensibly more advanced myoelectric prostheses, and that this is a much higher acceptance rate for myoelectric arms than the mere 5 percent of arm amputees in the general population who use them. For an advanced electronic arm to actually offer an improvement to the old-fashioned body-powered hooks, they will have to get better. A lot better. For DARPA program manager Ling, that would ultimately mean nothing less than going right to the source for full control of a full-featured prosthetic arm and hand: the brain.

Ling's Revolutionizing Prosthetics program built on the work of a program that predated him, DARPA's Human-Assisted Neural Devices program, or HAND, started by program managers Alan Rudolph and Eric Eisenstadt. The goal of HAND was to develop human-machine interfaces that read signals directly from the human brain to send commands to computers. Experiments with monkeys by DARPA researchers, most notably Miguel Nicolelis at Duke University and Andrew Schwartz at the University of Pittsburgh, showed that such interfaces were feasible.

This, in turn, showed that half of the problems facing the Revolutionizing Prosthetics program—direct brain control—could be solved.

The other half, however, were no less daunting.

To create a brain-controlled device was one thing, but to pack that device into the size and weight of a native human arm was quite another. Ling calls the human arm the most advanced biological tool in nature, and he may well be right. He asked me to picture all the things a human arm is required to do during the course of an ordinary day. It is strong enough to hoist a bowling ball, yet dexterous enough to pluck up a feather. It can play a piano, type on a keyboard, weed a garden, swim, shake hands, lift groceries, and in the case of a solider, field-strip an M16 rifle. It's impervious to water and to extremes in temperature ranging from below zero degrees Fahrenheit to more than one hundred degrees, and can operate all day long—as long, in fact, as its owner is awake and active. Nothing created by the most advanced science and engineering could come anywhere close to matching this performance. Ling asked me to think of the robots at Disneyland. "You go down and look at their robots and say, 'Man, look what they've got: hands that are moving and arms that are moving, and blah, blah, blah.' But what you don't see is what's behind the curtain. . . . It's this monstrous machine. Well, that just won't do."

Not if a machine is to be anything like a real replacement for a missing human arm. Such a machine has to be more than merely wearable. It has to be *comfortable* if a person is to wear it all day long, because, as Ling pointed out, "you don't want to put a fifty-pound thing on somebody— it'll tilt them over to one side."

So. To revolutionize prosthetics, first develop the means by which a human brain can control a machine. Then shrink an industrial robot to perhaps a hundredth of its size and weight, including its power source. Which power source has to be better than anything currently available.

"I mean," said Ling, "you don't want the thing conking out two hours in and you've got to plug yourself into a wall." This being DARPA, the first adopters would be wounded soldiers, who could then make the decision whether or not to go back to active duty. Soldiers such as the young man Ling treated in Baghdad would certainly head back into combat, which would be a good thing for the army, but, Ling told me, "Remember that in the world of America it's your choice to stay in the army or not."

Oh, and one other thing. "You're going to love this part," Ling told me. "We're not gonna sit around and wait for thirty years for something to pop out. . . . We want something to come out in four years." Four years—that is, by the end of 2009—to create a brain-controlled artificial arm that will do all or most of the things expected of a human arm, that looks like a native arm in order to allow its users to go about their daily lives as smoothly as possible, and that also provides its users with a sense of touch to make their lives even easier. "And that is very, very DARPA-hard," Ling said, summing up the goals of the project. "You would think that no maniac on earth would want to even attempt to do this."

IN AUGUST 2007, I checked in at the security desk in a nondescript, low-slung white building on the campus of Johns Hopkins University in Laurel, Maryland. The building was part of Johns Hopkins's Applied Physics Laboratory, or APL. Some of the maniacs attempting to build Ling's arm, it turned out, worked here. This was my first expedition into the mysterious and shadowy world of DARPA. No approval or clearance from DARPA headquarters needed, I had made direct contact with the team that was building Ling's arm and talked my way into an invitation to visit the lab.

Like DARPA, APL had been founded in response to a crisis, and like DARPA, it found renewed purpose after its immediate challenges had been met. APL was established at Johns Hopkins by the Office of Scientific Research and Development, or OSRD, in March 1942, just three months after the Japanese attack on Pearl Harbor catapulted the United

States into World War II. APL's initial mission: coordinate the top-secret effort to develop and manufacture the so-called VT, or proximity, fuses that eventually helped the Allies to victory with artillery shells that could automatically explode in proximity of a target rather than having to actually strike it.

APL became a DARPA contractor during the agency's first year of existence, 1958, developing the first satellite-based navigation system as part of a navy-sponsored program called Transit. APL developed the concept behind the Transit system—pinpointing the position of a receiver on Earth by measuring the Doppler shift of signals from orbiting satellites—as well as the satellites themselves for construction by RCA. Today, APL, still operating as a nonprofit research arm of Johns Hopkins University, plays host to hundreds of cutting-edge technology development projects, with an emphasis on national defense, but with a sizable civilian complement of projects as well, including spacecraft and spacecraft instruments development for NASA. Ling and DARPA director Tether picked APL out of a pool of at least a dozen applicants for the Revolutionizing Prosthetics assignment. APL then teamed up with a consortium of more than twenty other research institutions and private companies on what was to become a $100 million-plus project.

One OSRD official called the fledgling APL "the redball express leading from the smallest room of a research laboratory to the World War II battlefronts," and that description seemed nearly as apt in 2007 as it did in 1942. I first met the team in a cramped, windowless conference room. (The labs in which the engineers and managers developed the standards for and integrated the work of their partner institutions were similarly windowless and not much bigger.) Second in command for the project, John Bigelow, outlined the day's challenges. The former avionics engineer seemed unflaggingly cheerful about trying to motivate people to do the impossible.

Jon Kuniholm, the ex-marine, arrived late after a long night at the lab. He wore his body-powered hook. He had joined the team as part of his Ph.D. work in biomedical engineering at Duke University, one of the research institutions under APL, this one working on the fine controls needed to get the prosthetic hand to move through a series of preprogrammed grasps in response to muscle contractions by its user. It wasn't quite the brain-machine interface that Ling ultimately wanted to realize, but this refinement in the current state-of-the-art in myoelectric controls was an important step along the way.

As the only member of the team who actually used a prosthetic arm, Kuniholm had perhaps the strongest motivations among his colleagues for working to realize Ling's vision, but Bigelow, too, saw the project as a personal quest. He had been an engineer and manager in the defense industry, and was finding it hard to keep himself motivated when colleague Stuart Harshbarger, a fellow electronics engineer, recruited him to work on Revolutionizing Prosthetics. "It's been tremendously rewarding and exciting for me," Bigelow has since told me. In contrast to his feelings about working on warplane navigation and weapons targeting systems, "For me it's just a case of doing something that can give back."

Project leader Harshbarger, who joined the group working in the lab later in the day of my visit, views prosthetics as nothing less than his life's mission. As a boy he saw firsthand what the loss of a limb could do to a person's will to live. His grandfather had lost both of his feet to a lawn mower, and the young Harshbarger had watched the man's steady decline from that point on—not from his physical injuries, but because his handicap proved too great a challenge for him to overcome. Equally influential to the path Harshbarger's life later took was a neighbor, a Korean War veteran, who had lost an arm in combat, but who was so determined not to let his handicap get the better of him that he rigged a system of cables and pulleys that let him prune his own trees.

The team had been up late the night before trying to iron out the bugs for integrating the major components of the team's latest prototype, prosaically called Proto 2: the carbon-fiber sleeve with embedded electrodes, the control system, and the hand itself. At the moment, all the group's efforts were bent toward the team's first public demonstration, at the upcoming DARPA Technology Symposium, or DARPATech, DARPA's every-year-and-a-half technology showcase and recruiting event, coming up in just a week.

The big issue of the day was the so-called head stage, a microchip for interpreting control signals that was to be installed in a specially made square depression in the arm's carbon-fiber sleeve. Two other head stages had been fried by some unknown gremlin in the system. The team had just taken delivery of a third, but they were reluctant to risk it to the arm until they understood why the other two had burnt out. The head stage's job was to take signals from the electrodes in the sleeve and pass them to the processor for translation into movement.

Without the head stage, Kuniholm, the team's prime test subject, would have to stay attached to a bunch of external electrodes and bulky cables plugged into a box of electronics, where the signals from his arm muscles were collected and then forwarded to a card installed in a desktop PC. Software running on the computer translated the signals from the other gear into motion for a virtual reality hand standing in for the prosthetic still under construction in the lab. The head stage was designed to replace all of the intermediary components between hand and computer (which itself would be much reduced in size in an actual wearable prosthetic), and was crucial for restoring a future user's full mobility.

After discussing options for tackling the head stage problem (mainly, keep the new replacement head stage the hell away from Proto 2 until the glitch that was burning up head stages could be traced and eradicated), the team headed into the adjoining lab to resume the work they had put down the night before. It was here that the contributions of engineers at twenty-eight separate private companies and research centers based in half a dozen countries around the world came together under the direction of Harshbarger, Bigelow, and the APL engineers. The engineers had to successfully integrate computer software, specially fabricated microchips, tiny electrically powered motors, electrodes and other sensors, and a host of other systems in order to create Ling's Revolutionary Prosthetic.

For all that, the lab seemed rather modest in scale. It consisted of a cluster of smallish office spaces with the walls between them taken out. Workbenches cluttered with electronics test equipment, electronic parts, cables, wiring, oscilloscopes, computer workstations, and flat-panel screens lined the walls. A department store mannequin at the door wore the project's previous prototype, Proto 1, on its left shoulder. An adjacent workstation controlled its movements for testing. Cabinets in the main room were crammed with spare parts and tools, including a bunch of eerie-looking rubber hands and forearms, cutting tools, a drill, and a grinder for smoothing out connections on the electrodes in the sleeves of the prototype arms.

In another part of the lab, through an open doorway from the main room, one of the engineers began the painstaking work of dissecting tiny wires out of cables and peering at them through a microscope to hunt for short circuits or other anomalies that could have burned out the head stages. Behind him, Proto 2's upper arm graced the shoulder of a mannequin, a plastic green Mountain Dew bottle strapped to the forearm to replace the weight of the "cobot" (for "cooperative robot") that would go in the forearm and that was under construction at another workbench in the lab. As engineer Mike Bridges issued commands from his workstation beside the mannequin, the Proto 2 arm moved through a sequence of preprogrammed motions, including a swimming stroke, lifting its fingers to the mannequin's mouth as if to eat, and a crisp salute. A prominent yellow-and-red emergency Stop button sat near Bridge's right hand, ready to stop the arm in a hurry should it run wild. This was the main part of the demonstration that the team would take to DARPATech. It would be an impressive demo; the arm's motions were eerily human in their fluidity. Still, there was no mistaking the mechanical nature of the system. The whine of electric motors accompanied the robot's movements. And then there were those cables . . .

In a stark reminder of how far the project had to go in solving one of its major challenges, Proto 2 drew power from cables that ran along the back of the mannequin, down to the floor, and into an industrial-looking black box on the floor—part of the extensive external infrastructure that Ling wanted to shrink down and enclose within the confines of the arm itself. Conventional battery-and-electric-motor technology wouldn't get the job done, but one of APL's partners was working on a possible solution. Researchers at Vanderbilt University, in Nashville, had built a prototype system for actuating the arm not with electric motors but with hydrogen peroxide. The hydrogen peroxide, a full day's supply of which would come in a cartridge that a user could insert in his or her arm each morning, would flow through an iridium catalyst in response to the user's unspoken commands. The resulting reaction would generate steam in exactly the manner of hydrogen peroxide–powered rocket engines. In this case, the steam would drive tiny pneumatic actuators in the arm and hand. After it cooled, the steam would seep out through the arm's artificial skin, much like ordinary human sweat.

In the main part of the lab, other engineers bent over the forearm-shaped cobot and the shining metal bones and monofilament tendons that made up Proto 2's hand, making minute adjustments. Their immediate task was to integrate the two in preparation for attaching them to the rest of the Proto 2 arm. The cobot and hand represented an attempt to emulate the basic mechanics of the human hand. Electric motors in the cobot were designed to pull the white and yellow filaments that now trailed from the open wrist of the hand to make the fingers move. That hand, called the extrinsic hand because the motive power for pulling the tendons came from outside itself, followed the same basic design of the human hand, in which muscles in the human forearm pull tendons to move the fingers.

Just to cover all the bases, the engineers also were at work on a so-called intrinsic hand, which contained within it all the motors for operating the fingers. Even though they were leaving open the possibility that the intrinsic hand might turn out to be a better engineering choice than the extrinsic one, Harshbarger and his managers stopped short of suggesting that they were trying to improve on nature's design. All they were trying to do, Harshbarger insisted, was to restore as much of an amputee's native abilities as possible, in whatever way turned out to be the most efficient.

Kuniholm, for one, had no problem with the idea that he might be able to enhance his natural abilities with the machine that would eventually replace his missing right arm. When I remarked to him that it would be challenging to use a computer mouse with the Proto 2 hand and its fixed lateral position, Kuniholm shrugged. "Why do I need a mouse?" he said. "Why can't I plug my arm right into a USB port?"

He pulled up an office chair at a computer workstation and strapped the Proto 2's carbon-fiber sleeve, with its nest of attached data cables, onto his stump. On the screen before him, a computer-graphics display of a silvery metal prosthetic hand moved smoothly through pre-programmed gestures, including a rounded grasp, as though holding a ball, a more elongated rounded grasp, as though holding a cylinder, and touching the thumb to the forefinger. Each gesture was triggered by Kuniholm's flexing the residual muscles in his forearm. Electrodes picked up the electrical signals produced by the muscles and conveyed them via the cables and other gear to the computer, which interpreted them as commands for executing the particular series of motions Kuniholm wanted to make. Even as Kuniholm learned how to flex his remaining muscles in just the right way to produce the motions he wanted, the computer, too, was learning how to interpret those motions correctly. Both were striving for a perfect melding of man and machine.

In the near future, the APL team planned to improve on standard myoelectric control techniques by which the Proto 2's sleeve read Kuniholm's muscle contractions by injecting tiny devices developed at the Rehabilitation Institute of Chicago, or RIC, and dubbed injectable myoelectric sensors, or IMES, each about the size of a grain of rice, directly into the user's residual muscles. Ensconced in the muscles whose signals the engineers wanted to read, the IMES would pick up much cleaner signals, uncluttered by the noise commonly picked up by surface electrodes a wearer generated as he or she moved through the day. Since they were so much smaller than surface electrodes, clusters of IMES embedded in the same part of the user's body would also be able to much more cleanly distinguish among the competing signals of different muscles, giving them a higher resolution, and hence their user a finer degree of control over the prosthetic arm. The IMES would transmit their signals wirelessly to computer chips on the prosthetic.

The only way to begin to approach a natural arm's full dexterity, however, would be to surgically attach electrodes directly to the arm nerves, or even in the appropriate control centers of the brain. Although both options were under development by APL's collaborators, Kuniholm, for one, didn't relish the idea of having experimental electrodes installed in his head. For now, he would stick with noninvasive control systems, hence the need for the preprogrammed gestures. The surface electrodes lacked the resolution required for fine motor motions and could read only the grossest signals. Kuniholm could move the virtual hand on the screen through some basic moves just by thinking about them, but for finer movements, he had to learn combinations of bigger movers, which the computer translated into smaller ones. It was a process akin to using keyboard shortcuts to instruct a personal computer to execute a series of commands.

In spite of the long hours they had been at it, the team maintained an easy working style, bantering as they went about their tasks. "Talk to the hand!" Kuniholm cried at one point as he and his colleagues worked to try to bring the software controlling the hand and the computer executing Kuniholm's muscle contractions into closer communication.

Next stop, DARPATech, where the team would face some DARPA-sponsored competition. . . .

IN TYPICAL DARPA fashion, director Tony Tether had been reluctant to put all his bets on one performer for Revolutionizing Prosthetics. Ling had agreed, and thinking further about it, he said, "You know, not everyone might want to have these super-small electrodes put in their brain."

Sure, it worked on monkeys, but no one knew what the long-term effects of permanently installing wires in a person's brain might be, least of all a young soldier who'd had an arm blown off in combat and who might otherwise be unharmed and have a lifetime ahead of him. For those people, Kuniholm included—and to hedge their bets, in case the insanely ambitious four-year arm project didn't succeed—Ling and Tether came up with a two-year arm project. The goal of that project was to create what Ling called the strap-and-go arm. "You wake up in the morning, you put it on, and off you go," he explained to me. "It doesn't require hooking up to your brain or anything like that." No way such an arm could ever be as dexterous as one controlled by a direct link to the brain, but it could still be a lot better than the hooks most soldiers were stuck with. And without the burden of the challenge of wiring a prosthetic up to someone's brain, a two-year-arm contractor would have a big head start over the four-year team.

Not that the task would be easy; the specs for the strap-and-go arm were plenty ambitious all on their own. Like the four-year, APL-built arm, the two-year arm would have to be comfortable enough to wear all day long, and it would have to have the staying power required to keep going after strenuous activity without forcing its user to plug into a wall. One of Ling's requirements was that the arm be quiet, producing no more than fifty decibels during operation, because, Ling said, its wearer shouldn't have to worry about sounding like Robocop when watching a movie in a movie theater and disturbing other moviegoers. Tether and Ling ended up handing the job to DEKA Research and Development, a company with 250 employees based in New Hampshire and run by Dean Kamen, the inventor who gave the world the Segway Personal Transporter—that computerized two-wheel scooter for adults—in 2001.

The DEKA Research team shared a display booth with the APL Revolutionizing Prosthetics team at DARPATech in August 2007, and I, along with a gaggle of high-ranking military brass, got to see them side by side. Together, they stole the show.

The DEKA arm was modeled by double-arm amputee Chuck Hildreth. The forty-four-year-old had lost both of his arms at the shoulder when working a summer job twenty-six years before. While painting an electrical power substation, he had grabbed a power line he thought had been turned off, suffering severe burns in the process. He hadn't picked up a power drill in twenty-six years, he said, until he had strapped on the DEKA arm. He showed the conference goers how he could not only operate the drill, but also pick up small candies and transfer them easily to his mouth to eat. The system, a left arm in this case, was well on its way to fulfilling Ling's mission for the two-year arm—a prosthesis that would not require its user to submit to implants but that would nevertheless prove more useful than the decades-old hook designs still in such wide use.

Whereas the APL prototypes sought to contain all of their systems within the size and weight of an adult human arm, the DEKA arm suffered from no such constraints, giving Hildreth a somewhat menacing cyborg aspect. The arm's aluminum-and-black-plastic structure projected outward from Hildreth's left side noticeably farther than a native arm would, blending in appearance with the drill it wielded. In the space where Hildreth ended and the robotic arm began, I could see the atrophied stump of Hildreth's upper arm, all of a few inches long, deftly maneuvering to manipulate the specially constructed joystick strapped to it. Dispensing with myoelectric controls, the arm instead took its input from any available body part. Hildreth turned around to show me the cable running down from the power pack on his back into the back of his shoe, where it terminated in another specially designed joystick. Hildreth had had to learn combinations of commands activated by movements of his stump and by his right foot. This setup required a fair amount of training to allow its wearer to do useful work, though Hildreth said he was pretty good at operating it after only thirty hours of practice. "I can't imagine what I could do after more time," he said, obviously delighted with the device.

The arm featured a system to actively stabilize itself and keep itself properly attached to the user's body. The thing weighed nine pounds, a good two pounds heavier than a native arm, and offset as it was from the wearer's body, it could easily tire its wearer after only a short time, if it were just left to hang there throughout the day. So instead of just passively dangling off the end of the shoulder, the arm actually hugged itself to Hildreth's body using a system of six air bladders to mold its harness around his upper body. Like the system that gave the two-wheel Segway uncanny stability, the arm continually adjusted itself in response to its user's motions, working to redistribute its weight evenly over the user's upper body. Hildreth told me the arm was extremely comfortable to wear.

DEKA Research, which had cut its teeth on high-tech wheelchairs, planned to carry Ling's arm project beyond DARPA's commitment. Its engineers would keep refining the arm and then, when it was ready, put it on the market. It was a prime example of DARPA's long history of fostering innovations that took on lives of their own after getting an initial push from the agency and integrated themselves into the fabric of not just the military world, but civilian life as well.

In the Revolutionizing Prosthetics display area at DARPATech with the DEKA arm project was the APL project's Jon Kuniholm. The team never did get that head stage–frying glitch licked in time for DARPATech, and so Kuniholm was stuck wearing the same bulky nest of data cables and box of electronics along with the Proto 2's carbon-fiber sleeve tethering him to a computer workstation that I had seen him with the week before. The Proto 2 arm moved through its preprogrammed sequences on its mannequin nearby, while some of the other APL engineers I had met talked with conference attendees. Double amputee Jesse Sullivan wore APL's Proto 1 arm on his left side. On his right, Sullivan wore a body-powered hook. The Proto 1 arm and hand moved fluidly as Sullivan spoke, gesturing to punctuate his speech. While the DEKA arm seemed more generally useful, this one appeared more lifelike.

Sullivan's facility with his prosthetic arm was the result of an experiment he'd taken part in at the Rehabilitation Institute of Chicago five years previously. Upper-arm amputees such as Sullivan presented special challenges to the use of myoelectric controls. Since they had no residual arm muscles that could be read with surface electrodes, prosthetists had to resort to placing electrodes on the large, easy-to-read muscles of the amputee's back. The user flexed his back muscles when he wanted the arm to move. It wasn't a terribly intuitive process—the user had to get used to moving the muscles of his back when he really wanted to use his arm. So research scientist Todd Kuiken, at RIC, decided to do myoelectric controls one better. In 2002, Kuiken surgically transplanted nerves from Sullivan's left stump into his chest muscles in a process called re-enervation. After the operation, Sullivan's chest muscles twitched in response to signals from his transplanted arm nerves. Electrodes on the skin of Sullivan's chest could then register signals from his arm nerve-activated chest muscles rather than his back. Now instead of flexing an unrelated back muscle to move his prosthetic arm, he simply thought about moving his arm.

Although it seemed more capable, the big, powerful-looking DEKA arm couldn't come close to matching the Proto 1's more human aspect. DEKA's Hildreth had to think more about what he was doing making his movements seem more robotic. The DEKA arm, when it enclosed my own flesh-and-blood hand in its hard grip, unadorned with the rubber prosthesis that would eventually soften it, actually made me a bit nervous.

Each arm had its unique strengths and weaknesses, but which one was better? In answer to the unspoken question, Hildreth and Sullivan challenged each other to a mock arm-wresting match, much to the delight of onlookers. While Hildreth and Sullivan didn't dare actually pit their expensive and irreplaceable prototypes against each other, the competition that the two arm projects represented could produce real winners—that is, anyone who had the misfortune to lose an arm in their prime, whether in the service of their country or not. But only if the arms could also be made affordable enough for the people who needed them. Kuniholm put that requirement at the top of his list of important challenges to be met if his and his colleagues' work was to have a real impact. After all, the best prosthetic arm in the world might turn out to be next to useless if ordinary health insurance wouldn't cover it. "Even in volume," Kuniholm wrote in an unpublished paper he shared with me, "most highly articulated hands would likely cost more than ten times the cost of current myo-arms—a cost that many insurance companies already object to paying. As an engineer, I'm thrilled to be involved in the creation and testing of this latest batch of prosthetic concept cars. As a customer, my greatest fear is that of failure to complete the cycle of product development and actually bring it to market. Without a market, none of this means anything." Which is why Kuniholm's design firm's Open Prosthetics Project focused on improving that low-tech but far less expensive design, the body-powered hook. "I think we can generate far more societal benefit if we give away information than if we commercialized and sold the ideas," he says.

Actually, that statement comes surprisingly close to summing up much of DARPA's work: find an area of technology that could go a long way toward serving the needs of the country if improved but that wasn't getting much attention in the private sector, put some well-considered research and development money into it to get it on its feet, and then cut it loose.

It's what happens after it gets cut loose that worries Kuniholm and other engineers and managers working on DARPA projects. To help counter some of those concerns, DARPA director Tether made it his mission to ensure that as many as possible of the agency's projects had a clear pathway to implementation by its customers—even to the point of not funding projects beyond their preliminary phases until working relationships with those customers could be established. Walter Reed Medical Center had already committed itself to testing Ling's arms, which boded well for their future use by the people who needed them.

Of course this path from research and development to widespread distribution and use is only for the projects that I know about and that I *can* learn about. About half of DARPA's work is classified, off limits to any ordinary citizen, and follows a much more secretive path from laboratory to field. Only someone on the inside could know the full extent of what the agency was up to and how its projects were put to use, director Tether told me when I finally got through to him to research this book. Nevertheless, I was determined to get as deep as I could inside those smoked-glass-and-steel walls of the agency's Arlington, Virginia, headquarters, to at least go where no journalist had gone before. Revolutionizing Prosthetics and DARPATech gave me my first hint of what that glimpse inside might show me, and I was hungry for more. But first, to get some perspective, I needed to go back in time. Specifically, to a fall day in 1957 that shocked the world.

A SPECIAL-PROJECTS AGENCY

OCTOBER 4, 1957, Redstone Arsenal, Huntsville, Alabama. It was the end of a long day for Neil H. McElroy, the nation's new secretary of defense, and he was ready for a drink.

Just days away from his swearing in and fresh from his job as head of Procter and Gamble, McElroy had a lot of homework to do, including touring the nation's major defense installations. High on the list was the Army Ballistic Missile Agency, or ABMA, based here in Huntsville. McElroy was getting the red-carpet treatment from General John Medaris, the agency's no-nonsense commander and the nation's top missile man. Together McElroy and Medaris had spent the day tramping all over the arsenal's forty-thousand acres looking at test stands, assembly buildings, and firing bunkers, with their respective entourages of military brass and aides in tow, and now they were settling down in the officers' club for cocktails before dinner. For McElroy, the visit was simply one stop among many, part of a crash course in the requirements of his new job. Hell, he was just trying to keep his head above water.

For Medaris, however, the visit represented nothing less than a fight for the survival of his command. His mission was to show McElroy that the ABMA had the people, the experience, and the resources to put the United States firmly in the lead in the area of missile development. "We determined to give him our frank feelings," Medaris wrote later in his memoirs, "backed by facts and figures, as to our record for delivery of what we promised, when we promised, and for the money originally stated. We now hoped that with a fresh and uncommitted mind the Secretary-elect would grasp the significance of our story."

Little did Medaris know that he was about to receive a tremendous helping hand, at least in the short term, from the least likely source. Ultimately, what transpired that day would set in motion a series of events that would put Medaris's team on the fast track to outer space—to orbit and beyond. Unfortunately for Medaris, however, they would also spell doom for the ABMA.

But at the moment, Medaris was busy thinking of all the ways that the ABMA had been given short shrift by the outgoing secretary of defense, Charles Wilson. As Medaris later wrote, "Without knowing much about McElroy, we still could not shed a single tear over Mr. Wilson's departure. It was our strong feeling that his tenure had been characterized, to put it charitably, by a complete lack of imagination." Specifically, Medaris and his team felt that they had "consistently been pushed aside in favor of almost unlimited support for the glittering but unsubstantiated promises widely publicized by the Air Force and its contractors."

The word of the day was *missiles*. Missiles for delivering bombs, nuclear and otherwise, and missiles sent flying at the United States by her enemies. In the postwar world, characterized by a tense standoff between the United States and the Soviet Union, with each working feverishly to build up her nuclear weapons, missiles—who had them and who didn't—could well determine the fate of the world.

Medaris had under his command the best rocket scientists and engineers in the world, as far as he was concerned, including Dr. Wernher von Braun, Nazi Germany's former head of missile development. No one but the ABMA had the expertise needed to keep the edge in missile development, and Medaris found it exceedingly frustrating to have that fact continually buried under the politics of interservice rivalry.

The air force had split from the army only ten years before, but it now claimed dominion over all that flew any real distance, including ballistic missiles. The army, however, traditionally in charge of artillery, claimed missiles for its own.

And then there was the navy, whose Naval Research Laboratory had in August 1955 won the mission of launching America's first satellite with its Project Vanguard. The planned satellite launch was ostensibly for peaceful purposes; President Eisenhower had publicly put forth the national goal of launching a satellite during the upcoming International Geophysical Year, a series of scientific missions timed to coincide with the eleven-year sunspot cycle and running July 1, 1957, through December of 1958.

Be that as it may, Medaris had no illusions about the real value of rockets that could launch satellites, as he admitted in his memoirs. "To a soldier, the promise of trouble-free communication to all parts of the planet—to say nothing of keeping the enemy under constant surveillance—made preeminence in the space field absolutely essential."

This was quite apart from the idea of launching actual weapons. Communications and surveillance were key to waging war, and satellites could provide both in spades. "It was perfectly clear to me—and had been clear to von Braun for years—that the first nation to establish a permanent manned space station would have taken a giant step toward domination of the whole planet. And a satellite—any satellite—was the first step toward such a station." Equally inevitable to Medaris was the idea that access to space would naturally lead to combat in space. "We knew that wherever man has gone, on land or in the air or under the sea, sooner or later he's managed to get into a fight. We saw no reason to suppose that space would be any different."

The question of who would launch the first satellites and gain the high ground in space was, pure and simple, an issue of utmost importance to national security. The best team for getting the United States up there first should be chosen for the job, it was obvious that the ABMA had that team, and Medaris had made it his mission that October day in 1957 to make sure McElroy got the message.

The fact was that Vanguard was an entirely new design "with little experience and no developed components behind it." Since it had been designed from scratch expressly for launching a scientific satellite, Medaris surmised that it was "sufficiently dissociated from any weapons system so that it could be an unclassified project, and this was appealing to the civilian scientists." At issue here was the idea, originating from President Eisenhower himself, that America's public, high-profile space program should be seen as essentially a scientific endeavor rather than a strategic or tactical one.

Von Braun was more charitable than Medaris on the subject of Vanguard, predicting that it "would become a very useful tool for space research," but with an important qualification—that in order to realize this usefulness, it would have to have the inevitable bugs of a brand-new rocket worked out first.

Von Braun, who had been part of the proposal process before Medaris took command of his team, figured Vanguard had been chosen over the Redstone Arsenal because of promises that "their ultra high performance rocket could orbit a payload of 40 pounds, while the most we could guarantee, after stripping the Redstone of everything not essential for a pure research satellite launching, was a ten-pound load."

While von Braun steered clear of suggesting that the Vanguard's specs were outright fabrications, Medaris pulled no punches in his memoirs. "Originally budgeted at $20,000,000, the Vanguard was to become a $120,000,000 boondoggle that contributed little or nothing to progress in space, future weapons, or the international prestige of the United States," he fumed.

To Medaris, the attempt to separate civilian space exploration from military missile development made no sense. Vanguard, he said, was nothing more than a "very expensive toy." And, as he noted in light of events soon to ring down the curtain on the ABMA, "This was to be only the first down payment on this costly and ridiculous division of the indivisible."

Von Braun, on the other hand, had always had in mind the peaceful conquest of space, if such was possible. "Ever since boyhood I'd dreamed of building a rocket that could fly to the moon and beyond into inter-planetary space. That was the one goal I had while I was working on the V-2s in Germany; I kept praying for the day our missiles would be used to expand, rather than destroy, man's place in the universe."

The battle for control of America's embryonic space program hinged on the missiles at the Redstone Arsenal. While denied authorization for the satellite launch they knew they could pull off on short order, von Braun and his army bosses had begun prepping their Redstone rockets for satellite launch anyway, hoping that, given how far behind Project Vanguard was, they'd one day get that authorization. They did this under the guise of developing the Jupiter intermediate-range ballistic missile, or IRBM. Jupiter had a proposed range for lobbing explosives or nukes of suitably destructive power 1,500 miles, versus the Redstone's relatively paltry 200 miles. It was a new project, and it was getting priority for funds and launches.

Medaris was keenly aware of the propaganda value in demonstrating long-range missiles to the world as quickly as possible. "Since such a mis-sile was to be mainly a deterrent," he said later, "I felt that it [was] quite important for our potential enemy to know we had one . . . even though we also [knew] that the payload to be fired was quite small." In a bit of administrative sleight of hand, Medaris directed his engineers to launch their Redstones as far as possible with a stripped-down payload—as part of the Jupiter program. Medaris and von Braun's engineers would just stick nose cones belonging to Jupiter on the top of Redstone's first and second rocket stages. "Jupiter Cs, we called them," von Braun said later. "Officially they were to be used for the study of the reentry problem of the new Jupiter IRBM. We did *not* point out that, with minor modifications, they could also serve as satellite vehicles."

In 1956, von Braun and his crew of rocketeers had readied two modi-fied Redstones, which they prosaically named Missile 27 and Missile 29, for the ABMA's long-range launch demonstration. If successful, Medaris planned to trumpet the launch to the world. If not, well, it would be just another test of a ballistic missile.

The team flew Missile 27 in a cargo plane to Cape Canaveral in "A-1 condition for firing," as von Braun put it later. Unfortunately, it was not in similarly excellent condition on landing. The rocket's skin had buckled in the fast-rising atmospheric pressure of a quick landing forced on the plane's pilot by bad weather. Von Braun glumly concluded they'd have to trash the missile and go with their backup, but the program's quick-thinking structures engineer figured the team could just pump compressed air into the thing and it would pop back into shape. Turned out he was right, and, disaster averted, the team set up the rocket on the pad at the Cape for six weeks of exhaustive testing before launch.

September 20, 1956, is not a widely commemorated date in the history of space flight, but it could well have turned out otherwise. Because they were not authorized to launch a satellite, the team replaced the solid fuel in the rocket's fourth stage with sand, to achieve the right weight for testing the lower stages, thus turning what could have been a satellite launch into a mere ballistic missile test. That night, Medaris and von Braun and their crew fired Missile 27, and it performed beautifully. Medaris thrilled to the sight of the missile, bigger than any he and his crew had launched before, rising on a column of flame, "majestically," as he put it, and thundering off into the darkness. Minutes later, telemetry from the rocket's payload told the rocketeers that the missile had flown a perfect flight, rocketing to an astonishing 600 miles in altitude—well out of the atmosphere and into the realm of orbital flight—and splashing down in the Atlantic Ocean 3,300 miles away. Von Braun jumped up there in the firing room and shouted, "The old cucumber did it!"

"Having succeeded beyond our best hopes," Medaris planned to publicize the success as far and as wide as possible, and put the world on notice that the United States had the capability to launch long-range missiles, and thus satellites as well. But it was not to be. Keep a lid on it, came the official word from the highest levels of command. Weeks later, after a rocket magazine leaked the story, President Eisenhower finally relented and announced that the United States had launched a rocket an astounding distance. But the full implications of what that meant never penetrated the public consciousness.

Medaris was crushed. To have pulled off the greatest feat in rocketry and not be able to tell anyone about it was pure torture. He likened himself to the minister in the old joke who preaches vehemently against Sunday golf playing, but whose own greatest weakness is playing golf. In fact, he lives on a golf course. One Sunday morning dawns bright and sunny, and the fairway outside his door looks so wonderfully luscious that the preacher just can't resist taking a single shot. God frowns down on this puny mortal who dares to defy him and determines to make him pay dearly for succumbing to temptation. The preacher's shot sails straight and true, and bounces, rolls, and drops into a spectacular hole-in-one. After he has calmed down from the excitement of making the greatest shot of his life, the preacher realizes how God has punished him: he'll never be able to tell a soul what he's done.

"I can say this as a fact," von Braun reported later. "On the morning of September 21, 1956, we knew that with a little bit of luck we could put a satellite into space. Unfortunately no one asked us to do it. We had to sit still at Huntsville with the rest of our Jupiter Cs gathering dust."

This was the situation during McElroy's visit to Huntsville on October 4, 1957, and Medaris was good and ready to bend McElroy's ear about it. That evening, after his tour of the Redstone Arsenal, McElroy greeted a group of guests, including several high-profile Huntsville citizens and local officials, at the officers' club. He had just settled down to relax with Medaris and von Braun, probably with drink in hand, when one of Medaris's aides, a lieutenant named William Magill, burst into the room. "General," he blurted, "it has just been announced over the radio that the Russians have put up a successful satellite!"

For a moment, the party was struck dumb.

Then, as Medaris put it later, von Braun began jabbering animatedly, "as if he had suddenly been vaccinated with a Victrola needle" (in other words, jabbed hard in the rear with a dull metal spike). The elephant in the room, namely the fact that America's most capable rocket shop had been sidelined just when the country needed its unique skills the most, had suddenly reared its head—and sat on the guests. "We knew they were going to do it!" von Braun all but shouted at McElroy, according to Medaris. "Vanguard will never make it. We have the hardware on the shelf. For God's sake turn us loose and let us do something. We can put up a satellite in sixty days, Mr. McElroy! Just give us a green light and sixty days!"

Von Braun himself was a little more subdued in a later recounting of the episode. As he recalled, he said to McElroy, "If you go back to Washington tomorrow, Mr. Secretary, and find that all hell has broken loose, remember this. We can get a satellite up in sixty days."

Medaris thought of all the launch preparations that would have to be made before then, and said, "No, Wernher, ninety days."

"Okay," said von Braun. "Ninety it is." Still, he maintained his conviction that the work could easily be done in sixty days.

Whether the launch could occur in ninety, sixty, or even two days did nothing to alter the basic facts, however. The reality was that October 4, 1957, would go down in history as the day that the Soviet Union, not the United States, launched the world's first artificial satellite. That the United States could have earned that honor more than a year before mattered not at all. It was the Soviet Union's artificial moon, not America's, that even now was beeping and bleeping high overhead.

It weighed only a couple of hundred pounds, but it was more than enough. The greatest fear among many in the military establishment had come true: the Soviet Union had developed the means by which to sling nuclear bombs around the world and blast American cities to smithereens, and the United States could do nothing to respond. The Soviets had just fired the first shot of the cold war, and they had scored a direct hit.

THE WEEK FOLLOWING *SPUTNIK* constituted a "prolonged nightmare" for the Eisenhower administration, an Eisenhower advisor later recalled. A steady stream of advisors came and went from the president's office, and each "had a longer face than the one before."

Of course, there were the inevitable "I told you so's," most notably from Deputy Secretary of Defense for Research and Development Donald Quarles, who as early as May 1955 had impressed upon Eisenhower the importance of being the first to launch a satellite because "its unmistakable relationship to intercontinental ballistic missile technology might have important repercussions on the political determination of free world countries to resist Communist threats," as he said in his written recommendations. Those recommendations had been forwarded to the National Security Council by the president's special assistant, Nelson Rockefeller, with the even stronger admonition, "The stake of prestige that is involved makes this a race that we cannot afford to lose."

Of course, Quarles was also the guy who had ultimately been responsible for putting America's bets on Vanguard. And yet, he was unrepentant, bolstered by Eisenhower's continuing irritation over the fight for control of America's missile programs. "I was particularly annoyed," Eisenhower later said in his memoirs, "by a complaint made public by two Army officers over the earlier decision to continue the Navy's Vanguard as the United States satellite program, when, according to them, a booster developed by the Army could have long since done the job." Never mind that the complaint might have been valid—the fact was that the services had no business airing their dirty laundry in public.

So it was that on the morning of October 8, even as Quarles was forced to admit in Eisenhower's office that the army could have launched the first satellite a year earlier, he successfully defended the choice of Vanguard over Redstone as America's first satellite launcher. The ballistic missile programs and the satellite launch effort, Quarles said in that meeting, ought to be kept separate. The scientific data to be gathered by Vanguard's purely scientific International Geophysical Year satellite was to be shared with all nations equally, Quarles reminded Eisenhower, and so it was right to give Vanguard priority. It was a point of view that Eisenhower heartily endorsed. "The American space program," the president was able to insist later, "in no way began as a race or contest with any other nation. Instead, the undertaking and all information gained from it were something of a gift to the scientific community of the entire world."

Redstone, by contrast, whose main mission was to deliver lethal payloads, was a closed project whose "many defense secrets would be jealously guarded." To deliver the best possible scientific benefit from a satellite launch, "and to keep the satellite effort from interfering with the high-priority work on ballistic missiles," Eisenhower believed, "it seemed mandatory to separate the two programs." Never mind that it was the navy that would provide the launch vehicle and facilities.

In agreeing with Eisenhower, Quarles was so bold as to suggest in his meeting with him, "The Russians have in fact done us a good turn, unintentionally, in establishing the concept of freedom of international space." The satellite was, after all, at this moment, flying over the airspace of countries all over the globe, thus asserting not that the Soviet Union had dominion over outer space, but that no one country had that claim.

Be that as it may, prudence was clearly called for at this juncture. That morning, Eisenhower quietly instructed outgoing secretary of defense Charles Wilson to tell the army to get the Redstone rockets ready for satellite launch "as a backup for the Navy Vanguard."

And what of the fear that the Soviet Union, in beating the United States to the punch in space, had also perfected the technology of ballistic missiles? Well, Eisenhower asked his advisors, what of it? Is there any truth to the notion that the nation is any more at risk today than yesterday? Eisenhower's chief of staff, Sherman Adams, put the question to the National Science Foundation president Detlev Bronk thusly: "Is there anything in the Soviet achievement to make us alter our research and development program, particularly in the missile field?"

"No," said Dr. Bronk, "in my opinion, there is not. We can't always go changing our program in reaction to everything the Russians do."

Thus satisfied that *Sputnik* did not constitute an immediate threat,

Eisenhower held a press conference on October 9 in which he attempted to quell the fears of the American people. He congratulated the Soviets on their achievement and said that the United States *could* have been first in putting up a satellite, but that doing so would have been "to the detriment of scientific goals and military progress."

The reporters attending the press conference didn't buy it, and they wasted no time in pelting Eisenhower with pointed questions about the Soviet's satellite technology and America's lack thereof.

"Mr. President," asked the *Portland Press Herald*'s May Craig, "Do you not think that it has immense significance, the satellite, immense significance in surveillance of other countries, and leading to space platforms which could be used for rockets?"

"Not at this time, no," the president replied. "No. There is no—there is—suddenly all America seems to become scientists, and I am hearing many, many ideas." He supposed that someday satellites would be able to transmit some kind of useful imagery of the ground to their operators, but that day was "a long ways off" and that we had nothing to fear from "one small ball in the air." Certainly, he added, "in view of the real scientific character of our development, there didn't seem to be a reason for just trying to grow hysterical about it."

"Mr. President," said NBC's Hazel Markel, pressing the point, "are you saying that at this time that with the Russian satellite whirling about the world, you are not more concerned, nor overly concerned about our nation's security?"

"Well, I think I have time and again emphasized my concern about the nation's security," the president said. "Now, as far as the satellite itself is concerned, that does not raise my apprehensions, not one iota."

What Eisenhower had failed to anticipate was that in a cold war, the "psychological advantage in world politics" he spoke almost offhandedly of in his press conference was every bit as important as technology and strategy.

But he was a quick study; he would soon learn that the better part of containing the growing national hysteria over *Sputnik* would have to include a very public push to advance American space and missile technologies. The two were inextricably entwined. Still, that wouldn't stop him from continuing to try to make a clear separation between them, including attempting to divorce America's future space program once and for all from the military machineries that spawned it. It was a move that would have major ramifications for decades to come.

Part of Eisenhower's reluctance to treat *Sputnik* as a significant military threat might be explained in part by his deepening distrust of what he was later famously to term the "military-industrial complex."

In his 1953 speech in which he enumerated the cost in human terms of spending a large portion of a nation's resources on its military assets, Eisenhower invited the Soviet Union to join with the United States in stepping back from the brink of militarization. "The subject was universal disarmament," he explained later in his memoirs. "The fruit of success," he said in his speech, "would present the world with the greatest task, and the greatest opportunity of all." The task would be nothing less than "a declared total war, not upon any human enemy, but upon the brute forces of poverty and need."

Eisenhower had yet one more reason for calm. The fact was he *knew* that the Soviets did not have the upper hand in missiles and the technology to build them. Since 1956, the top-secret U-2 spy plane, flying at an unassailable seventy thousand feet over the Soviet Union, had brought home photographic evidence of Soviet armaments. The photos revealed that the supposed missile gap, as it came to be known in the media, between the United States and the Soviets simply did not exist. Of course, Eisenhower couldn't very well make that assertion publicly. So he was forced to come off to his critics as hopelessly out of touch with the new dangers of the cold war.

MEANWHILE, the man who would be called upon to define the details of America's response to the perceived increased threat from the Soviets had problems of his own to attend to. On the morning of October 9,

1957, the fifty-three-year-old Neil H. McElroy entered his office on the third floor of Ring E in the Pentagon. He flipped on a red light outside the door to let his staff know that he was not to be disturbed, sat down at the U.S. government's largest desk, nine feet long by nearly five feet wide, and containing no fewer than twenty drawers, ordered the first of twelve cups of coffee he would drink that day, and got down to work as America's brand-new secretary of defense.

He had to work fast to get up to speed on his new job, but that suited him perfectly. The man, in fact, seemed to live his entire life in perpetual fast-forward. "You can't call a walk with Mac a stroll," a friend once said of him. "It's more like a run." McElroy's impatience with wasted time often flared into irritation, earning him a reputation as something of a hothead. He had a weakness for gambling and bourbon and, said a January 1958 *Time* magazine profile, "can use four-letter language that does not spell TIDE," one of his former company's major products. He had, in short, exactly the kind of mix of confidence, bravado, and drive, as well as experience, needed to take over the world's largest nonprofit organization in a time of crisis.

Eisenhower had first become impressed with McElroy's ability to get the job done, and get it done fast, in 1950, when the former was president of Columbia University. Eisenhower was looking for an extra $25,000 to fund a planned conference series focused on public issues. He pitched the project at a luncheon attended by a group of Cincinnati philanthropists and business leaders, McElroy among them. Intrigued by the project, McElroy approached Eisenhower afterward and told him to "wait around a few moments while I nail this thing together." Right then and there, he corralled the necessary donations from the people present. When Eisenhower became president of the United States, he tapped McElroy to head the newly formed White House Conference on Education in 1954. A year and a half later, McElroy and his team of experts—who included the president of the Massachusetts Institute of Technology, or MIT, James R. Killian, Jr., who was later also to have a prominent role in the Eisenhower White House—put together a no-holds-barred report blasting the current state of American education and recommending that the United States double its education spending.

Under McElroy, Procter and Gamble had twice earned honors as America's best-managed company. McElroy became known for having a knack for cutting through the noise of facts, figures, and other data to find the one error that needed fixing. He lived, breathed, and slept facts, according to his colleagues at Procter and Gamble, and it was this ability to quickly understand a situation and efficiently home in on the one relevant problem in a sea of competing data points that was to serve him so well as the Pentagon's new chief.

McElroy's hell-bent-for-leather attitude toward his job had another very important source: he was on borrowed time from day one. When he agreed to take the position, he told Eisenhower that he would do so on one condition: that he take a leave of absence from Procter and Gamble rather than quitting outright. The understanding was that the job of secretary of defense would be merely a brief interlude, only two years, in his private-sector career. As with many successful people in industry going into public service, he was reluctant to give up entirely the private-sector salary and all the corresponding benefits, but at the same time felt compelled to answer the call to serve his country.

This built-in term limit may have contributed greatly to McElroy's success as secretary of defense. He wouldn't have time to become entrenched in the world's largest bureaucracy. Nor would he be distracted by trying to further his own career in government, because he knew that in a short time he wouldn't have one.

McElroy's insistence on limiting his time in office was also to set an important precedent for the research-and-development agency he was soon to be instrumental in creating. It helped to keep him, along with the new agency, clear of the interservice rivalries that so irked Eisenhower and that had now become McElroy's concern. For instance, one of the pressing problems on McElroy's desk was what to do about all the missiles.

True to form, McElroy got to work cutting through the clutter to find

the crux of the matter, which was this: the United States needed intermediate-range ballistic missiles, also known as IRBMs, with a range of 1,500
miles or so, as well as intercontinental ballistic missiles, or ICBMS, and it
needed them as soon as possible to act as a deterrent to the Soviet Union's
build-up of same. The competing missile programs of the various services
represented needless wasted effort and hence wasted time.

Key to all of this was the nation's research-and-development capability. Detlev Bronk, head of the National Science Foundation, once
remarked that the president "liked to think of himself as one of us." It
was natural for Eisenhower to turn to the Science Advisory Committee,
established by President Truman as part of the Office of Defense Mo-
bilization, for advice in a time of crisis. The committee included some
of the keenest minds in the country—physicists, mathematicians, and
electrical and nuclear engineers. On October 15, the president called in
the committee to get their read on what *Sputnik* might mean to national
security.

In their meeting, the scientists confirmed Eisenhower's view that
the nation's missile programs were in good shape—on par with or better than the Soviet technology. No dramatic moves were needed there.
Their real concern was what would happen in five or ten years. Was the
United States devoting sufficient resources to science and technology
development to keep its edge? Or would it steadily lose ground in the
face of relentless progress by Soviet scientists born into a culture that
treated science and engineering as a "kind of social passion." That kind
of passion for problem-solving was "missing in American life," committee chairman Isidor Rabi lamented. "When I was growing up, all the
boys wanted to play first base. Now most of them seem content to sit in
the bleachers."

Eisenhower wasn't so sure about that. "I questioned the assumption
that the Russians were trying to inspire all their people to enter scientific
pursuits," he mused later. "I thought instead, from watching their record
over the years, that they had adopted a practice of culling out their best
minds and ruthlessly spurning the rest, so far as higher education was

concerned." Eisenhower told the committee that he'd do what he could to inspire the nation's young people to get excited about science and technology. "The people," he said, "were alarmed and thinking about science and education," and that should be enough, with just gentle prodding, to keep kids inspired all on their own.

Turned out he was right. As a *Sputnik*-induced wave of rocket mania swept the nation's youth, boys everywhere began scraping the heads off matches, collecting ammonium nitrate from fertilizer, zinc dust, sulfur, and all matter of other "hair-raising chemicals," as a *Time* magazine article in May 1958 put it, filling lengths of pipe with homemade brews of solid rocket fuel, and enthusiastically blowing their concoctions skyward, often to detrimental effect.

Still, Rabi did have one suggestion that Eisenhower took to heart. Calls in the press and in Congress for the creation of a top-level "missile czar" had been growing ever more shrill. Perhaps, Rabi suggested, it was time to appoint a science advisor who would directly brief the president and his staff on scientific matters of national urgency.

Eisenhower took a dim view of the whole idea of a "czar," but he thought the idea of a science advisor was terrific. "Neil had better hear this right away," the president told an aide.

McElroy was at his desk when one of three telephones in his office jangled. It was the White House phone, and that meant the boss was calling. The call gave him only a few minutes warning of the limo that brought the members of the Science Advisory Committee, sent by the president. The scientists repeated for McElroy the essence of the meeting they'd just had at the White House, and McElroy had no choice but to drop everything and listen. Science and scientists were getting their due in the Eisenhower White House. And this was only the beginning. Soon after those meetings, Eisenhower appointed a top-level science advisor. Only a month had passed since the *Sputnik* launch.

The man the press just couldn't resist dubbing America's "missile czar" was James Killian, president of MIT and a member of the education committee that McElroy had headed up three years earlier. Killian moved into an office in the Executive Office Building, across the street from the White House, where he could have immediate access to the president.

"I bring not a sense of alarm or despair," Killian assured a reporter not long after his appointment, "but rather a lively sense of urgency."

Secretary of Defense McElroy, meanwhile, had ideas of his own. He well understood the importance of research and development to any large, modern organization. As president of Procter and Gamble, he was downright fanatical about R and D, fostering the development of new products at a breakneck pace, so that during his last year as president, before he became secretary of defense, fully 70 percent of the company's profits came from products that had been invented just within the previous twelve years.

It seemed to McElroy that R and D could save the day for the government, too. That is, if it could be uncoupled from the politicking and bureaucracy that had hobbled America's missile and space efforts so far.

On November 6, 1957—three days after the Soviet Union launched *Sputnik II*, a half-ton behemoth with a live dog aboard—McElroy fired off a memo to his legal advisor, Robert Dechert.

"Does the Secretary of Defense," McElroy wanted to know, "have legal authority ... to establish, in his office under an Assistant to the Secretary, a new unit which would centralize control . . . of anti-missile weapons and satellites?" And could he fund it with money already allocated to the armed services without going back to Congress to ask for separate funds? Finally, to make maximum use of those funds with minimal bureaucratic headache, could this hypothetical new organization "call on the military departments to perform administrative functions such as the preparation and signing of contracts, disbursement of funds, and other support activities?"

No problem, Dechert replied in short order. All he'd have to do would be to notify the chairmen of the two Armed Services Committees.

Then make it happen, McElroy told Dechert.

On November 15, as the pressure on President Eisenhower ratcheted up, he took a break at his favorite retreat at the Augusta National Golf

Club, in Georgia, where normally he could relax a bit, enjoy a few rounds of golf, and carry on some business in his office over the pro shop. But this time things were different, and he worried out loud to an aide: "I'm not sure that I'll get too much out of this stay. Too much on my mind."

By November 20, McElroy was ready to make his move. During his first testimony before Congress as secretary of defense, he announced, "We plan to establish in the Department of Defense a special agency to handle our satellite and space research and development projects. Tentatively we are thinking of calling this—and this has not been announced—the Special Projects Agency of the Department of Defense." This special agency, McElroy told members of the House Subcommittee on Department of Defense Appropriations, might also oversee America's antiballistic missile research. And what was to become the future agency's salvation, "Other projects and programs may also be assigned from time to time."

The new agency would be staffed by both civilians and military personnel as the secretary saw fit. It would become the research-and-development branch of the military, with creative-minded program managers in charge of funding out-there research around the country that might or might not bear fruit, but that had a potential for big pay-offs. Those projects that did succeed would be handed off to the service branches of the military for further development and eventual use.

"The vast weapon systems of the future," McElroy told Congress, "in our judgment need to be the responsibility of a separate part of the Defense Department . . . in order to follow these various will-of-the-wisps—if they are originally in that kind of state—and carry them through to a point where there can at least be a determination of their feasibility and what their probable cost might be." Making clear that this small, nimble organization—answerable directly to the secretary of defense—should not be limited to space and missiles, McElroy told his audience in summing up his plans, "We are thinking of this Special Projects Agency as having a function that extends beyond the immediate foreseeable weapons systems of the current or near future."

Eisenhower loved the idea, and so did Killian, the new science advisor. Eisenhower had never made any secret about his disgust with the interservice rivalries that resulted in what he saw as a ridiculous and very expensive duplication of efforts at the air force, navy, and army. In fact, as early as 1945, years before becoming president, he had told cadets at West Point that he'd love to see all the services in one uniform. "I have always believed," he clarified in his memoirs, "that a nation's defense would be most efficiently conducted by a single administrative service, comprising elements of land, sea, and air. . . . Successful defense cannot be conducted under a debating society."

McElroy's Special Projects Agency seemed to Eisenhower just what the doctor ordered. Ultimately, he had in mind an ambitious, all-encompassing civilian space agency that would provide the focus for the nation's efforts to dominate the Soviets in space. But such a large-scale bureaucracy as that would take time to set up and fund. What he needed was something he could launch into action right *now*. Secretary of Defense McElroy had come up with what Eisenhower thought of as an elegant stopgap solution. The services hated the whole idea of an independent R-and-D organization within the Department of Defense, of course; they saw it as yet another challenge in a long line of such challenges to their jurisdictions that had begun in 1947, with the creation of the Department of Defense itself.

Eisenhower returned to Washington from his retreat on November 21 "looking drawn," as one reporter noted. He seemed "nervous and grumpy" during a Cabinet meeting on November 22, and then he shut himself up in his office and worked until late into the night on a speech addressing *Sputnik* that he was to deliver in Cleveland on the twenty-sixth.

Meanwhile, it had taken all of two days for the armed services and the Joint Chiefs of Staff to fire back their official responses to McElroy's plan. They started off on a condescending note, and went downhill from there. "The Air Force appreciates that the subject proposals are suggestions and implementation of Presidential policies for better and improved directional management in relation to certain areas of research and development," stated the secretary of the air force, James Douglas. By relegating the plan to establish the Special Projects Agency to the status of a mere proposal, Douglas made it clear that he had no intention of swallowing it whole.

The navy and air force took a slightly more conciliatory tone in working to try to contain the new agency's purview to just satellites and antimissile work. They suggested in their responses that all language to "other" projects be cut from the agency's charter. The navy went so far as to suggest changing the name of the agency to the "Space Vehicle and Ballistic Missile Defense Agency," to make its limitations absolutely clear.

All three services and the Joint Chiefs also pushed for removing the words *operate*, *operation*, and *management*, as such words implied an independent agency calling its own shots as an entity distinct from the services and therefore competing with them for funds and big new programs. The new agency, the services said, should not have its own laboratories and other R-and-D facilities, but rather should always farm those functions out through the services.

Killian put perhaps most succinctly the skirmishing that developed over the creation of the new Pentagon R-and-D agency: "It is strange now," he said in his memoirs from the vantage point of twenty years later, "to recall the fantasies that Sputnik inspired in the minds of many able military officers. It cast a spell that caused otherwise rational commanders really to become romantic about space. No sir, they were not going to fight the next war with the weapons of the last war; the world was going to be controlled from the high ground of space. . . . And they were convinced that their service, be it army or air force, was best qualified to develop the exotic technology that would be needed for space warfare—and for civilian use, too." Not surprisingly, Killian singled out Major General Medaris of the ABMA in Huntsville for special comment, saying that his and von Braun's efforts to give the army the major role in space were characterized by a "fierce religious zeal."

Secretary McElroy gave little quarter. For him the situation couldn't be plainer. To do its job, his new agency had to have the autonomy needed to develop the programs it saw fit, and only after it had demonstrated the feasibility of those programs—when they were good and ready, in other words—would it hand them off to the appropriate services. This, McElroy was convinced, was the only way to get the work of building those "vast weapon systems of the future" accomplished as efficiently as possible.

The president himself wasn't able to give McElroy much support at the beginning of the struggle over the new agency. On November 25, he suffered a stroke. It was the third time in as many years that he had been felled by illness. Luckily the stroke appeared to affect only the speech center of his brain, giving him difficulty only in finding the right words to go with his apparently unimpaired thoughts. Still, Eisenhower himself was alarmed enough to consider resigning if he wasn't satisfied with his progress toward improved health.

On November 29, McElroy circulated a revised charter for DARPA to the armed services and the Joint Chiefs with only minor modifications. By then the agency had acquired the name under which it would start life: the Advanced Research Projects Agency, or ARPA. McElroy had promised Congress that he'd do something about the name that threatened to add one too many "Special" organizations to an already crowded pantheon that included the Office of Special Operations and the Armed Forces Special Weapons Project. Special or not, the name had to encompass enough, McElroy testified to the House Appropriations Subcommittee on Defense Appropriations during hearings on ballistic missiles, "so that the tent can cover additional projects as they come along and give evidence of some future potential." No, he most definitely did not want to pin himself down to something like the navy's suggestion: the Space Vehicle and Ballistic Missile Defense Agency.

On December 6, 1957, a month after the launch of *Sputnik II*, Project Vanguard was finally ready to launch America's first entry into the space race. The twenty-three-pound International Geophysical Year satellite wasn't planned for launch until March, but this test flight, of a three-pound, six-inch-diameter test satellite, was rushed into the breach as America's answer to the *Sputnik*s.

Unlike the previous, undisclosed Soviet launches, many of which had ended in failure, America's first attempt to launch a satellite went on in full public and international view. Millions around the world watched on newsreels as the missile majestically rose from its stand at Cape Canaveral. It reached an altitude of all of four feet before its single liquid-fueled, thirty-thousand-pound-thrust, first-stage engine abruptly quit, and the seventy-two-foot missile collapsed back to Earth. The nose cone containing the test satellite toppled off the rocket like a gigantic dunce cap as the missile's propellant tanks exploded on impact in a spectacular fireball. The next day, the press ridiculed Vanguard variously as Puffnik, Kaputnik, Stayputnik, and perhaps most ingloriously of all, Flopnik.

His mood by no means improved, by December 13, Eisenhower nevertheless felt well enough to attend a North Atlantic Treaty Organization conference in Paris, thinking of this as the ultimate test of his ability to lead the nation. Fortunately the trip went off without a hitch, and afterward Eisenhower was able to set about putting down the rising military opposition to the creation of his new agency. On January 7, 1958, he submitted a request to Congress to fund the agency with an initial $10 million for the 1958 fiscal year. Then he publicly spanked the generals in his State of the Union address two days later for their "mistaken zeal in promoting particular doctrine," saying that while he didn't intend to spend his address berating them for the internal skirmishing, he didn't intend to tolerate it, either.

"I am not attempting today to pass judgment on the charge of harmful service rivalries," he said. "But one thing is sure. Whatever they are, America wants them stopped." To which sympathetic members of Congress wildly applauded. Then, after mentioning that "in recognition of the need for single control in some of our most advanced development projects, the Secretary of Defense has already decided to concentrate into one organization all the anti-missile and satellite technology undertaken within the Department of Defense," he made it clear that he expected that the secretary's directives would be honored. "Another requirement of military organization is a clear subordination of the military services to duly constituted civilian authority," he said. "This control must be real; not merely on the surface."

Eisenhower also used the State of the Union message to reassure the American public. The Strategic Air Command and the navy, he said, gave the United States more than enough nuclear firepower to annihilate any country that dared to launch a nuclear strike at the United States. "Even if we assume a surprise attack on our bases, with a marked reduction in our striking power, our bombers would immediately be on their way in sufficient strength to accomplish this mission of retaliation. Every informed government knows this. It is no secret." Indeed, the nation's survival might well *depend* on this not being a secret. *Sputnik* did nothing to upset the deadly game of teeter-totter between the United States and the Soviet Union that came to be known popularly as mutual assured destruction, since no rational head of state could dare launch a strike against the United States.

With the military challenge to the creation of McElroy's new R-and-D agency thus headed off at the pass, McElroy and Eisenhower were free to deal with fresh grumblings from Congress, led by House Armed Services Committee chair Carl Vinson, on whether the president and his secretary were overstepping their authority in creating a brand new agency. Vinson, a Democrat from Georgia who had been entrenched in the House of Representatives since 1914, had been affectionately known by some in the Eisenhower administration as "Uncle Carl" because of his typically strong support for the military, but he didn't like Eisenhower and McElroy sidestepping Congress to create ARPA.

After some serious haggling, McElroy and his team finally agreed with the legislators on some language that would go in an authorization bill for ARPA. The bill would give the secretary the authority to do with his funds as he saw fit, without specifically mentioning the Advanced Defense Research Projects Agency.

On January 29, 1958, the United States finally got its first satellite into orbit, thanks to von Braun, Medaris, and company at the Army Ballistic Missile Agency. The eighteen-pound *Explorer 1* boosted into orbit atop its Jupiter-C (actually Redstone) rocket to "shouting, singing, and cheering," as von Braun later reported.

McElroy's office at the Department of Defense issued its order establishing ARPA on February 7, 1958. It directed the new agency, under a to-be-appointed director, to assume responsibility for "advanced projects in the field of research and development as the Secretary of Defense shall, from time to time, designate by individual project or category."

The directive authorized the new agency not only to manage its research projects at the facilities of the individual armed services but also to contract with other government agencies and, giving it the broadest possible range of control, "to enter into contracts and agreements with individuals, private business entities, educational, research or scientific institutions including federal or state institutions." Finally, the directive gave ARPA the authority to "acquire or construct such research, development and test facilities and equipment as may be approved by the Secretary of Defense, in accordance with applicable statutes" (to smooth any still-ruffled feathers in Congress).

Sufficiently mollified, Congress passed the law authorizing the creation of ARPA on February 12, 1958. Instead of naming ARPA directly, it gave the secretary of defense the authority to "engage in such advanced projects essential to the Defense Department's responsibilities in the field of basic and applied research and development which pertain to weapons systems and military requirements as the Secretary of Defense may determine after consideration with the Joint Chiefs of Staff."

The law's failure to mention the new agency explicitly was to make some of the agency's future leaders nervous. Without specific authorization, could the agency even be said to have a legitimate existence? Could it simply be allowed to vanish at any time? So tenuous was the new agency's footing that one new recruit, for the agency's top administrative position, no less, wouldn't start work until he had exacted a promise from the man hiring him that there'd be another job for him if ARPA "went up in blue smoke."

But the lack of specificity suited others just fine. "Out of sight, out of mind" was the way they felt about it. It would help them stay below the radar when the inevitable skirmishing started up again. One clause did cause them real concern, however. It gave the secretary of defense authority over advanced space projects handed down from the president only "for a period of one year from the effective date of this Act."

ARPA, said Senator Lyndon Johnson, "was 'a temporary expediency'— a means of keeping the momentum alive in our space and missile programs without foreclosing the future."

It was a warning that although ARPA had survived its difficult birth, it remained on life support, kept alive only to address the immediate crisis. Eisenhower had grander plans for the nation's space program. And they didn't include ARPA.

SECRETARY OF DEFENSE McElroy was uncharacteristically slow in naming his selection for the head of ARPA. "I consider this to be one of the key appointments I will ever make," he said before his announcement, "and I don't want to be rushed into it."

It wasn't until ARPA actually opened its doors at the Pentagon that McElroy picked a General Electric vice president named Roy Johnson to head the agency. Johnson had no prior experience in government, and like McElroy, he was determined to do his duty and get out as soon as reasonably possible. He was happiest when "building something," he said. After he got something in motion, he said, "I lose interest." That described the nature of ARPA itself—then and in the future—perfectly. Johnson figured he'd stick around a year and a half to two years—long enough to get the new agency on its feet, and then head back to his high-paying civilian job. ARPA represented a tremendous cut in pay for Johnson—from $158,000 a year, no mean sum in 1958, to a paltry $18,000.

Johnson saw launching satellites as his main mission, and he set about assembling the right team for getting the job done. Not that he himself had any engineering expertise, though he was to become known for having a keen interest in ARPA's space mission—so much so that he was later to be called "a missionary for space, a zealot," in an official ARPA history. In a January 1959 speech, Johnson referred to America's role in space as an endeavor "so potentially magnificent it can only be upstaged by the goal it strives to attain—a goal that involves not only a just and lasting peace here on earth, but, in a deeper sense, the moral extension of man's dominion in the universe."

Heady words for a manager by vocation who in fact aspired to be a painter. Johnson's office at GE in New York was decorated with his own paintings, done in a modernist style that he self-deprecatingly referred to as indicative of "early Johnsons." When the fifty-two-year-old retired in the not-too-distant future, he planned to attend art school.

Johnson's first order of business was to hire a chief scientist, someone who could make the kind of educated decisions that were out of Johnson's realm of experience. Wernher von Braun looked like an obvious choice, but when Johnson discovered that von Braun wasn't going anywhere without the cadre of German rocket scientists who had stayed by his side since the war, he looked elsewhere. Instead, Herbert York of Livermore Labs became ARPA's first chief scientist.

Next came a discussion among Johnson, York, and McElroy about who would turn the bold ideas generated at ARPA into actual working hardware. Since McElroy had fought so hard to give the agency the authority to run its own labs, an obvious solution would be to take over one or more existing facilities and go from there. The Army Ballistic Missile Agency, whose German rocket scientists had built the Redstone rocket, and the Jet Propulsion Laboratory that had built *Explorer 1*, for instance, could just keep doing their work under the ARPA rubric, as could the Naval Research Laboratory, which had been in charge of the *Vanguard* launch. The men ultimately decided against having ARPA run its own labs, however, for fear that the sheer administrative weight of operating those big shops would drag the agency down. ARPA was supposed to get things done in a hurry, after all. Thus, its preferred mode of operation would be to work directly with the scientists and engineers it needed on a contractual basis, wherever they happened to be employed.

Now Johnson's job became much easier. All he had to do was hire a much smaller group of engineers and scientists to dream up good projects for ARPA and then dole out contract money to the people at those external research labs to carry them through.

To streamline operations even further, Johnson, McElroy, and York decided to contract out the agency's program managers, too. They worked out an arrangement whereby an independent think tank called the Institute for Defense Analysis, or IDA, would lend out scientists to the new agency, including people hired specifically at ARPA's request. With the beginnings of his staff in place, Johnson got right down to the work he saw as key to national security—developing and launching satellites.

On March 24, 1958, President Eisenhower approved ARPA's first space projects. But he qualified that approval by saying, "I do so with the understanding that when and if a civilian space agency is created, these projects will be subject to review to determine which would be under the cognizance of the Department of Defense and which under the cognizance of the new agency."

Only two days later, Eisenhower announced that he would submit legislation to create a new agency that would take control of America's civilian space programs. Underlining what he saw as the importance of the civilian role in space, he released a report written by the President's Science Advisory Committee, chaired by James Killian, that recommended emphasizing the civilian and peaceful use of space over military use. Then, on April 2, Eisenhower submitted the bill to create the National Aeronautics and Space Administration, NASA, from the existing National Advisory Committee for Aeronautics, or NACA. And he asked the head of NACA and Secretary of Defense McElroy which DOD programs should go to the new civilian agency.

Finally, just to cover his bases in space and other advanced research,

Eisenhower recommended creating the position "director of defense research and engineering," above the ARPA director, to oversee all DOD R and D.

Roy Johnson was livid. As far as he was concerned, Eisenhower was cutting ARPA's legs off even as it was trying to get on its feet. Johnson took to Capitol Hill to argue strenuously that ARPA's role as the military's space agency should be preserved even in the presence of a civilian agency. "The legislation setting up a civilian group should not be so worded that it may be construed to mean that the military uses of space are to be limited by a civilian agency," he said. "This could be disastrous." Particularly irksome to Johnson was the idea that ARPA might have to be answerable to a civilian agency. "For example," he said, "if the DOD decides it to be militarily desirable to [create a] program for putting man into space, it should not have to justify this activity to this civilian agency."

Johnson's pleas fell on deaf ears, and NASA took over the space mission of ARPA and, indeed, for the most part, that of the entire military itself. For example, unlike ARPA, it had no qualms about acquiring the Army Ballistic Missile Agency. Medaris then lost his command, and von Braun stayed on at the facility that came to be known as NASA's Marshall Space Flight Center.

Though ARPA's tenure as America's space agency was extremely brief, the agency nevertheless managed to leave its mark. Its space legacy includes starting work on what was to become the world's most powerful operational rocket.

To realize the military's vision for manned space flight, or "man-in-space" as the idea was known at the Pentagon, Johnson figured he needed a super-thrust rocket, one that could deliver a then-unheard-of 1.0 to 1.5 million pounds of thrust. Johnson knew full well that the Eisenhower administration would balk at such an obvious ploy to extend the military's manned reach into space, so he floated the new rocket as a launcher for communications satellites that could weigh up to a whopping ten thousand pounds each. The idea flew. ARPA program managers David Young and Richard Canright, both IDA employees, developed the idea of ganging together seven, eight, or even nine IRBMs to create ARPA's super-rocket, then took the idea to von Braun's team at Huntsville. Von Braun and his crew had themselves studied the idea of clustering four rockets together to create a bigger-thrust rocket, but the idea of putting at least eight of them together really got von Braun fired up, since a rocket with eight engines could conceivably survive the loss of any one of them in flight and still accomplish its mission.

The rocketeers in Huntsville got to work on what became known as the Saturn IB rocket, with ARPA funding, in the summer of 1958, just around the time NASA was getting established. Once ensconced at the fledgling civilian space agency, NASA managers seized on the Saturn IB as the crown jewel of the ABMA when making their decision to turn the army's missile development operation into NASA's Marshall Space Flight Center. It was a Saturn rocket—the 7.5-million-pound-thrust Saturn V, powered by five clustered F-1 engines (which also started life as an ARPA program)—that sent Americans to the moon from 1968 to 1972, finally beating the Russians in space. The performance of the Saturn V has never been matched.

ARPA's contribution to space firsts also included the series of satellites that comprised Transit, the navy's satellite navigation system also known as NAVSAT, which was designed at Johns Hopkins University's Applied Physics Laboratory and which paved the way for modern satellite-based navigation systems. The satellites, operational in the early 1960s, allowed navy submarines to navigate by measuring the Doppler shifts of radio signals sent by the satellites. The Doppler data enabled computers on the subs to determine the positions of speeding satellites and thus triangulate their own locations. The state-of-the-art computers only just barely fit through the hatches of the submarines, and only then by having their cases rounded off. Nevertheless, the system was a resounding success, remaining operational until the more advanced GPS—with its atomic-clock-enabled time signals providing more accurate fixes than the Doppler-based system—made it obsolete in the 1990s. ARPA more than proved its worth as a pioneer in space research, even as that research mission was taken away from it.

The creation of the office of the director of defense research and engineering, or DDR&E, in 1958—a position between ARPA and the secretary of defense in the Pentagon hierarchy—became another challenge to ARPA's existence. ARPA chief scientist York became the DDR&E, and it was he who got most of ARPA's space programs—including the Saturn rocket—transferred to NASA. The remaining space programs—military satellites—York sent to the air force. It was the beginning of the end for Johnson's tenure as ARPA director. "Saturn," as one of Johnson's managers at ARPA put it later, "was the biggest [personal] Roy Johnson contribution and he did it over the dead, dying and bleeding bodies of just about everybody."

After presiding over ARPA while it had its guts, its most important work, torn out of it, Johnson just couldn't maintain his enthusiasm for his job. He took to showing up at the office a lot less frequently and delegated responsibility to his subordinates a lot more often. He left in November 1959, having served three months shy of the two years he had expected to stay in the job. In one of his last acts as director, he instructed his staff to figure out what role ARPA could have, if any, in the absence of a space mission. The resulting paper explored several options for the future of ARPA, including abolition, but ultimately recommended that it redefine its mission. The year 1959 was a time of tremendous upheaval at the Pentagon, not just at ARPA. Almost at the same time as Johnson, Secretary of State McElroy announced his own return to private industry, and in May, Deputy Secretary Donald Quarles—he who had advocated Vanguard as America's first satellite launcher over Explorer—died of a heart attack.

It was by no means certain that ARPA would survive the 1950s. "The first stepchild of the space age seems rapidly on its way out the nearest and most convenient exit," editorialized Martin Caidin in the December 1959 issue of *Data* magazine, "convenient being merely the covering phraseology for official embarrassment at having a dead cat hanging in the fruit closet."

The decision was York's to make. Had ARPA served its purpose, he wondered, or did it yet have more life in it? Ultimately, he decided that the organization that he, McElroy, and Johnson had built could bring its special abilities to bear on technical and scientific problems outside the realm of space. As York put it later, "There is just less argument with ARPA about getting [advanced research] going, and doing it, than there is with the Services. The Services, whenever the budgets get tight, they squeeze out those things which there is nobody in the Service with passion for."

Equally fortunate for ARPA's continued survival was McElroy's early insistence that its purview not be limited to space and missiles. Accordingly, the agency had run other programs, even under Johnson's watch. These included research into materials science, a DARPA-invented term that describes the now-well-established study of building structures for specialized uses, such as to withstand extreme heat. DARPA effectively created the field by bringing together the separate disciplines of metallurgy, ceramics, chemistry, and even, in recent times, biology, to study how all of these different materials could interact to create novel structures. The impetus for doing this was the need for very strong materials that could survive the launch and reentry of spaceships and boosters but that nevertheless were light enough to let those vehicles get off the ground. Leo Christodoulou, DARPA's deputy director of the Defense Sciences Office in the late 2000s, described his office's approach to problems of materials science like this: "In the past, the researcher confronted with a challenging application would look at a stock of possible materials and ask, 'Which one can I use?' Our materials researchers use a different approach. We ask, 'What do we need [in order] to make a material for a particular application?' There is no bias toward what was used in the past." One early success was the development of nickel alloys called super alloys, which enabled high-speed turbines to survive the punishing heat of sustained jet engine combustion, and were used extensively in the construction of the X-15 winged spaceships flown by NASA and the air force in the 1960s.

ARPA also put significant resources into antiballistic missile and nuclear test detection research. Its work in nuclear test detection, in particular, proved enormously influential, making possible the ratification of the first nuclear test-ban treaties agreed to by the United States and the Soviet Union. With satellites funded by ARPA's nuclear test monitoring project, Vela (Spanish for "watch"), ready for launch, President Kennedy could sign the Limited Nuclear Test Ban Treaty of 1963 secure in the knowledge that the United States had the ability to detect the telltale optical and electromagnetic signatures of Soviet nuclear tests.

Satellites could monitor nuclear tests that took place in space or in the atmosphere, but they couldn't peer underground. That's where Project Vela's seismic monitoring stations came in. The problem of detecting underground explosions caused by nuclear bombs was more difficult than simply watching for radiation from orbit. Sensors appropriately placed at good vantage points around the world would not only have to detect the faint tremors caused by distant blasts, but also sort through myriad signals from natural earthquakes to find the unique seismic signatures of nuclear explosions.

The result was the World Wide Standardized Seismograph Network, or WWSSN, a network of 130 monitoring stations built to uniform standards in order to provide the most accurate correlation of data. To gather and sift through all that data, ARPA created the Vela Seismological Center in Alexandria, Virginia, which became the agency's biggest and most advanced computer center. The ARPANET became the means by which data from around the world was sent into the system. "We didn't really intend to wire the world together in the Internet," Steve Lukasik, ARPA's director from 1970 to 1975, reflected later, "but that's what happened." More specifically, Lukasik said, "we wired the seismic world together. Otherwise, it used to be you would mail in your tracings every day or two and it would take three weeks before all the data came in." To analyze the data, ARPA funded research that led to advances in artificial intelligence, neural networks, and other innovations in information technology that continue to pay dividends today.

At least one DARPA manager has gone so far as to credit his agency and the researchers it funded as part of Project Vela with essentially creating, as was the case with materials science, an entirely new scientific discipline. Before Project Vela, says Ralph Alewine, who directed DARPA's Nuclear Monitoring Research Office from 1980 to 1986, "the field of seismology did not exist." While that may be something of a stretch, since scientists had studied earthquakes with seismographs since at least the late 1800s, it is true that before Vela, there was no truly coordinated international network of such seismographs feeding data into one computer network for analysis. And that has made all the difference in enabling scientists to confirm what had before been only theorized about, and to develop an unprecedented understanding of the way the planet Earth is structured. Regardless of whether seismology existed as a field before Vela, scientists lacked the most basic information about what caused earthquakes. "Without the DARPA system," says Alewine, "they would not have had the data needed to discover plate tectonics." If scientists ever do grasp the Holy Grail of seismology, not only detecting but actually predicting earthquakes, it will be because of DARPA's groundwork.

To get all those monitoring stations of the WWSSN in place required an unusually close collaboration with an international community of scientists and government agencies, which, says Steve Bratt, DARPA program manager for data processing from 1993 to 1996, "created a high level of trust in what DARPA was doing." As a result, "I think people were largely grateful the U.S. was willing to fund so much of this research that improved global capabilities"—capabilities such as monitoring and gathering data on tsunamis and earthquakes around the world.

By every measure, ARPA did indeed bounce back from the loss of what was initially its primary mission as America's first space agency. Starting with a few core non-space research areas, ARPA was able to greatly expand its purview, most famously in the area of information technology.

THE INTERGALACTIC
COMPUTER NETWORK

IN THE POST-NASA WORLD, ARPA emerged as a sort of dumping ground for military programs that could find no other home. In 1961, the air force needed to unload what one former ARPA staffer called an expensive white elephant (in the form of a major piece of computer hardware called the AN/FSQ-32XD1A), and the fledgling R-and-D agency ended up with it.

The 250-ton machine had been built by IBM as a spare for the air force's Semi-Automatic Ground Environment, or SAGE, program. The program featured twenty-three IBM-built AN/FSQs scattered around North America in hardened bunkers, each designed to interpret radar data about incoming Soviet bombers and use that information to coordinate launches of American interceptor missiles. The air force had no idea what to do with the spare AN/FSQ-32XD1A, which also came with a roomful of technical support from a RAND corporate subsidiary called Systems Development Corporation, or SDC, in Santa Monica, California. So the office of the DDR&E decided to palm the thing off, tech support and all, on ARPA. As much as $50.8 million for shipping and handling later, ARPA found itself with its very own computer.

In those days, when state-of-the-art computers cost tens of millions of dollars, filled entire floors of buildings, and required their own full-time support staff, just owning one was a big deal. It was definitely a major asset for ARPA. The question was what the hell to do with it.

The DDR&E had the idea that the computer could support research into command and control—that is, the coordination of orders delivered from commanding officers to the field—and "to provide a better understanding of organizational, informational, and man-machine relationships." Beyond that altogether vague description, no one really knew to what use to put it. Mainly its transfer to ARPA got it off the air force's books and also provided some work to an important contractor, SDC, at a time when it was idle between major projects.

ARPA's director at the time, Jack Ruina, had his hands full with more pressing projects, most notably those concerning ballistic missile defense and nuclear test detection, so he appointed J. C. R. ("Lick") Licklider to run programs associated with the new computer along with some new behavioral sciences projects. "Jack seemed too busy," Licklider recalled later. "He was just relieved to get somebody to run the office. . . . I talked with him periodically [and] he would make suggestions about directions of things, but pretty much let me do what I wanted to do."

Licklider might have seemed to some an odd choice for the first director of ARPA's new computer project office, which acquired the name of the Information Processing Techniques Office, or IPTO. He wasn't an electrical engineer or a software engineer by training. He was, in fact, a psychologist, most recently employed by Bolt, Beranek, and Newman, or BBN, a Cambridge, Massachusetts, acoustic engineering firm. Licklider was an expert in the rarefied field of psychoacoustics, the study of how people perceive sounds. But he was also extremely interested in computers, not just for the brute mathematical processing power for which they were commonly used in the day, but as tools for enhancing the cognitive abilities of human beings.

Before going to work for BBN, Licklider had worked on the SAGE project, trying to figure out the easiest, most intuitive way for human operators to control the ungainly computers and the big radar screens at their consoles. These were the so-called human factors of computer system design, and to Licklider, they were every bit as important as all the other elements that went into a computer system. All the processing power in the world would do little good if a human operator couldn't effectively make use of it. Working on SAGE got Licklider thinking about the future possibilities of big computer systems. Instead of correlating radar tracking data, could such a system be made to access libraries of text? Of raw scientific data? What about pictures? A system like that could actually be seen as an extension of the human mind rather than the mere calculating tool, the glorified abacus that computers were then mainly used as.

Licklider formally laid out his ideas for melding human and machine intelligence in a visionary 1960 paper called "Man-Machine Symbiosis." "As has been said in various ways," Licklider wrote in his paper, "men are noisy, narrow-band devices, but their nervous systems have very many parallel and simultaneously active channels. Relative to men, computing machines are very fast and very accurate, but they are constrained to perform only one or a few elementary operations at a time." There must be a way, Licklider reasoned, to meld the best thinking powers of both people and machines into a symbiotic whole that "will think as no human brain has ever thought and process data in a way not approached by the information-handling machines we know today."

When Licklider went to ARPA, those ideas became the heart of the agency's computer research efforts, extending all the way through the decades into the twenty-first century. Licklider even saw beyond the kind of human-computer interaction he fostered at ARPA to a day when computers would actually outthink humans in a new field of artificial intelligence, in which "electronic or chemical 'machines' will outdo the human brain in most functions we now consider exclusively within its province." Some of DARPA's advanced computing projects are now dedicated to just that pursuit.

Essentially, Licklider wanted to develop the new field of interactive computing. Today the systems by which humans interface with their computers are considered crucial to the success of a computer system as a whole, with major corporations such as Microsoft and Apple building their fortunes around them. But in 1961, the idea of investing major research money in the study of how computers and their operators could better interact was a radical one. Licklider was a maverick who believed, along with just a few of his colleagues, "that people could really become very much more effective in their thinking and decision-making, if they had the support of a computer system, good displays and so forth, good databases, computation at your command. Of course that wasn't believed by people in the command and control field."

If ever there was a job for ARPA, this was it—cutting-edge research in a field that was only just being recognized, and that wasn't yet fully appreciated by the armed services. When he came to work at ARPA, Licklider laid out his proposed research areas to director Ruina. Although he had no clearer idea than anyone else of where interactive computing could lead, Ruina saw the potential of Licklider's idea at once. "The hardware was there," Ruina knew, "but what to do with it was clearly lacking—what to do with this tremendous power." He was very much on Licklider's wavelength about the "use of computers for other than purely numerical scientific calculations. It impressed me," he said later, "as being something that was important."

Happily ensconced in his new office at ARPA headquarters in the Pentagon, Licklider went about laying the groundwork for what was to become one of ARPA's most influential lines of research. One of his first moves was to expand his program beyond the one computer ARPA had at its disposal. He wanted a community of computer experts and their machines at his disposal. "Essentially what I did on the command-and-control thing," Licklider said later, "was to try to figure out where the best academic computer centers were, and then go systematically about trying to get research contracts set up with them, aiming for three or four major ones and then a lot of little ones." In a 1963 memo, he affectionately referred to the expanding cadre of computer experts and machines he brought together under the ARPA umbrella as "the members and affiliates of the Intergalactic Computer Network."

One of the first research organizations Licklider tapped to help realize his vision of man-machine symbiosis was the Stanford Research

Institute, or SRI, in Menlo Park, California, at the time a part of Stanford University. Licklider found in SRI computer researcher Doug Engelbart a fellow visionary. Engelbart had summarized his vision of the future of computers in a proposal to the Air Force Office of Scientific Research as early as December 1960, under the heading "Augmented Human Intellect." The air force gave Engelbart and SRI money to study his ideas, but it was ARPA, at Licklider's direction, that gave him the funding to actually start working with hardware and software. Engelbart's project seemed like a perfect use for that surplus air force computer ARPA had inherited, and in 1963, Licklider gave SRI remote access to it. Engelbart's group could now log on and do their research on the part of the ARPA project Licklider called machine-aided cognition.

Remote access to a computer represented the state of the art in computing, but working with thinking machines at arm's length like this was not at all what Engelbart had in mind. He envisioned a much more personal relationship between people and computers than was possible to realize without having direct access to the computer on which he did his research. To do what he wanted to do, which included such farsighted experiments as creating "devices that enable the user to interact effectively with a displaying computer for document generation and modification faster than a typewriter can achieve" (i.e., the then-novel concept of word processing), Engelbart would need a dedicated machine all his own with which to tinker and attach his own custom-made hardware. That wasn't an easy proposition in the 1960s, and it would have to wait.

By the time Licklider left the agency in 1964, ARPA's IPTO oversaw a collection of computer laboratories that extended across the country, including at MIT; SDC, home of ARPA's first computer; the University of California at Los Angeles, or UCLA; and SRI. These ARPA-funded research centers and others maintained some of the most sophisticated computers of their time, along with the heavy-duty brain power to run them. ARPA-funded computer research was now making breakthroughs at a rate unmatched anywhere else in the world. As MIT artificial intelligence pioneer Marvin Minsky noted in a 1975 interview, ARPA's work in computer science "probably contributed more than all the other computer science laboratories in the country in the most advanced things." In fact, he said, "it's almost impossible to think of anything important that has happened anywhere else." One of the most important ARPA-funded developments came in the area of time sharing, the technology that allowed SRI's Engelbart and his team to remotely access the ARPA computer.

In the beginning if you wanted to use one of the bit-processing powerhouses in Licklider's Intergalactic Computer Network to run some research program, you had to go to it—actually fly across the country if necessary. Since these machines filled entire rooms, they couldn't very well be brought to the people who needed them for their research. One of Licklider's legacies at ARPA was to help work out ways to allow users to share time on a computer so that they could work on their programs at the same time, and also work remotely. The users did this by interacting with the computers through remote terminals. The computer would parcel out memory and processing cycles according to how many users and their programs happened to be working on it at a given moment. Time sharing, as this computing technique came to be called, brought Licklider and his colleagues much closer to the dream of truly interactive computing. "This new way of using computers," ARPA's annual report for 1966 explained, "allows an individual with a problem to use the computer in a conversational, interactive way, thereby permitting new approaches to solving a variety of problems that heretofore were not amenable to the computer."

A former NASA program manager named Bob Taylor rose through the ranks of IPTO program managers to become the office's third director in 1966. He later described the time-sharing setup at his disposal. "In my office in the Pentagon, I had one terminal that connected to a time-sharing system at U.C. Berkeley. I had one that connected to a time-sharing system at the System Development Corporation, in Santa Monica. There was another terminal that connected to the Rand Corporation." Taylor could interact, long-distance, with any one of those computers through the remote terminals in his office.

He himself saw the limitations of interacting at a distance with a computer, however, and the year he took over IPTO, he got the agency to spring for a brand-new SDS 940, the first computer designed expressly for time sharing, for Engelbart's group at SRI. Even before then, Engelbart and his team had been at work building monitors (most computer output in those days came through Teletypes and punch cards), input devices, and software to allow users to interact with computers as never before. But the new computer, along with ongoing support from ARPA, NASA, and the Air Force Office of Scientific Research, allowed them to rocket to a whole new level, culminating in a system they called simply the oN-Line System, or NLS. It was the precursor of virtually every aspect of what we know today as the personal computer.

By December 1968, Engelbart and his team at SRI were ready to reveal the fruits of their labor to the rest of the world for the first time. The occasion was the Fall Joint Computer Conference at the San Francisco Civic Auditorium, and the place was packed with, by one account, two thousand to three thousand computer researchers. The event would go down in computer lore as the Mother of All Demos, and it would forever change the direction of computer hardware and software development—though many of the concepts Engelbart introduced would take decades to come into their own.

In a staid, matter-of-fact delivery that made it all seem so *inevitable*, Engelbart proceeded to demonstrate for the first time, from his seat at a workstation whose display was projected overhead for all to see (itself no easy feat—he'd had to borrow the only projector on the West Coast capable of projecting a video image from NASA), word processing, video chat (with colleagues at their own workstations back at SRI forty miles down the coast), real-time collaborative document editing, hypertext links that connected key words to relevant supplemental material, and, among all the rest, the computer mouse. "I don't know why we call it a mouse," Engelbart monotoned. "Sometimes I apologize. It started that way, and we never did change it." To say that the demo was a tour de force was an understatement. In a world of punch cards, Teletypes, and batch processing, and in which computers were anything but personal, the debut of the concepts behind today's personal computers blew the minds of Engelbart's audience. At the conclusion of the demo, they erupted to their feet in a standing ovation. "I looked up," Engelbart later recalled with wonder, "and everyone was standing, cheering like crazy." One conference attendee later described the demo as hands down the most astounding event he'd ever witnessed. Another topped that by saying that it not only altered his concept for what computers could do for people, but also changed his *life*.

Surely missed by most in the audience at the demonstration, amid all the astonishing innovations, was a brief mention of a new project, "this ARPA computer network—experimental network that's going to come into being in its first form in about a year and end up sometime later with some twenty experimental computers in a network." Little did Engelbart's audience know that that experimental computer network would turn out to be just as world-changing as anything they had just witnessed.

In his 1963 "Intergalactic Computer Network" memo, Licklider had warned of the problems inherent in isolated computer systems. He foresaw a day when continued advancements in the field would depend on the ability to link computers together so that "at least four large computers, perhaps six or eight small computers" could all be brought together to share their "disk files and magnetic tape units" and all of their remote terminals and Teletype stations to create a more powerful, flexible, and ultimately, more complete whole.

In 1966, IPTO's Bob Taylor could readily see the need for a more comprehensive computer system. Each computer that Taylor could interact with in his office spoke its own language, required a different log-in procedure on a different terminal, and, of course, couldn't talk to the others.

What he needed, thought Taylor, was a way to use just one terminal and one log-in to access all three of the computers he had access to. In other words, he needed a way to get the computers to talk to one another. A computer network. And one that linked computers not only of the same type, as had already been done experimentally, but of different designs. Once he had that, the computer scientists around the country working on DARPA-funded projects could have it, too. They'd then be able to share computer resources and double, triple, even quadruple their computing power. With a computer network through which any computer could communicate with any other computer on the network, researchers could make much more efficient use of then-still-very-expensive computer resources—they could all get more time on more computers.

Taylor described his brainstorm later as "kind of an 'Aha' idea." Right then and there, he headed over to the Pentagon's E-Ring to the office of Charlie Herzfeld, ARPA's then-director, and pitched the idea to him. Herzfeld listened, asked a few questions, and then, as Taylor described it later, "he pretty much instantly made a budget change within his agency and took a million dollars away from one of his other offices and gave it to me to get started." Back at his own office in the D-Ring, Taylor looked at his watch and let out a breath. "Jesus Christ," he said to himself. "That took only twenty minutes."

The computer network program became known as ARPANET, and it was to be the first network linking computers of different types. Taylor had pitched it to Herzfeld in simple terms, with a confidence that belied the extreme difficulty in pulling off this technical challenge—the kind of challenge that future DARPA program managers would term DARPA-hard.

In those days even time-sharing computer systems were few and far between, with only high-level researchers interested in using them. These were unique machines, or at least in very limited production runs, manufactured for specific purposes, with software written expressly for them. Each machine had its cadre of technicians and programmers who communed with it in a process that seemed to draw more from the black arts of the occult than the realm of engineering. There were no software manuals, no standardized operating systems. Each machine had its own input/output system, and there certainly was no standard-ized means of transmitting information from one machine to another. To create ARPANET, Taylor and the people he hired would have to cre-ate all of those things. They would have to, in fact, lay the foundations for what we know as the modern computer industry today.

Step one for Taylor was to hire a program manager for ARPANET. Larry Roberts, a twentysomething computer expert specializing in the field of computer graphics at MIT's Lincoln Laboratory, home of Project SAGE, turned out to be the perfect fit for the job. Taylor had already funded a Roberts-run networking experiment during which Roberts and his team had successfully linked two computers thousands of miles apart. That made Roberts one of the few people in the world actually qualified to advance the field of computer networking.

The only problem was, Roberts liked his job at MIT too much to leave it to become what he considered a mere Washington bureaucrat. Taylor spent an entire year working on Roberts, all the while not finding a suitable substitute. Finally, in desperation, he looked into the Lincoln Lab's sources of funding, found that it got more than half from ARPA, and thereby got the leverage he needed to pry Roberts loose. At Taylor's request, ARPA director Herzfeld called Roberts's boss at Lincoln Labs and explained how it would be in the lab's best interest to let Roberts come work with him. That did the trick, and though Roberts went to ARPA grudgingly, he quickly immersed himself in his new work.

To solve the problem of getting such disparate computer systems to talk to one another, ARPA researchers hit upon the solution of building the ARPANET out of identical go-between computers whose only job would be to interpret data from the host computers and pass the data, properly translated, on to the network. Each host computer would be coupled to its own go-between computer. The job of building the go-betweens, refrigerator-size interface message processors, or IMPs, went to BBN.

The IMPs were also known as packet switches because of the scheme they used to pass on data. To protect against data loss associated with transmitting data over long distances, and to maximize the limited bandwidth available to them over the 1960s telephone network, the ARPANET's engineers borrowed a technique proposed first by Leonard Kleinrock in his MIT Ph.D. dissertation, and then independently by RAND Corporation engineer Paul Baran and Donald Davies in Britain. Baran arrived at the scheme as a way for the U.S. communications network to survive a nuclear attack. Instead of taking a direct line of communication from a message's source through a centralized distribution point to its destination, as the U.S. phone network was designed to do, he proposed chopping each message into discrete chunks he called message blocks, routing each chunk through the communications network via different pathways, and reassembling them at their destination. "I get credit for a lot of things I didn't do," Baran said in a 2008 interview. "I just did a little piece on packet switching and I get blamed for the whole goddamned Internet, you know?"

BBN shipped the first IMP to UCLA in September 1969. There, a team of faculty members and students attached it to their mainframe computer. SRI got an IMP of its own in October and connected it to its computer, and on October 29, 1969, the first two nodes on the ARPANET (and the Internet it birthed) communicated with each other up and down the California coast over four hundred miles of leased telephone lines. In a glitch presaging the annoyance of Internet users everywhere when experiencing dropped connections, the SRI machine accepted exactly two letters from the UCLA computer in that first connection before crashing.

As Leonard Kleinrock, the faculty member in charge of the project at UCLA, was later to note, that one small step for a computer and one giant leap for computerkind went completely unnoticed by the world at large. He said later, "1969 was quite a year. Man on the moon. Woodstock. Mets won the World Series. Charles Manson starts killing these people here in Los Angeles. And the Internet was born. Well, the first four everybody knew about. Nobody knew about the Internet." It would be decades before the significance of the achievement was widely recognized, and even today, the full potential of the globe-spanning network and the millions of nodes that it spawned is only just being realized.

After they had accomplished the feat of connecting disparate computer systems into a single network, ARPANET researchers turned to the problem of connecting different *networks* together into one apparently seamless whole. By 1972—the year ARPA acquired the *D* for *Defense*, to become DARPA—Roberts had succeeded Taylor as DARPA's head of IPTO, and he tapped BBN's Bob Kahn to solve the problem of tying together networks of computers that relied on different communications protocols. The radio and satellite networks DARPA was by then experimenting with for the armed forces didn't play well with the more stable telephone networks used by the ARPANET. Radio and satellite transmissions kept fading in and out as their transmitters passed in and out of range. They simply couldn't operate using the same data-transmission protocols used by the land-based ARPANET, with its steady, always-on (or usually on, as the case may be) connections. So Kahn decided to split radio and satellite networks into their own separate networks, which would then communicate with the ARPANET as separate entities in themselves. The computers on the new networks would get their own IMPs that spoke their own languages, and each network would get its own network-level IMP-like machine called a router, which would in turn translate between the individual networks and the broader collection of networks to be called the Internet.

Kahn turned to a group at Stanford University led by Vinton ("Vint") Cerf, who had been a graduate student on the team that installed the first IMP at UCLA, to work out the details of how the constituent networks would communicate. Kahn and Cerf called the process of tying collections of computer networks together "inter-networking." It took a full year for Cerf's group to develop the rules by which the individual networks would communicate with one another. They called these rules the transmission control protocol, or TCP, and these are some of the same standards by which today's Internet operates. Cerf and Kahn published the first paper describing TCP in 1974, and thus secured their place in history as two of the main progenitors of the Internet. But the honor of making the first true Internet transmission—that is, across disparate computer networks—went to researchers at SRI.

On August 27, 1976, a fine sunny day, the team pulled a van crammed full of computers, radios, and other electronic gear up to one of their favorite watering holes, a bar and grill called the Alpine Beer Garden—or Rossotti's, by its regulars—in nearby Portola Valley. They ran some cable from the van (affectionately dubbed the Bread Truck), stationed in the gravel parking lot, to the wooden picnic tables out back, where they could celebrate their achievement with beer and burgers. There they set up a remote terminal and fired off their weekly report to DARPA.

The big, clunky typewriter-like terminal tethered to a van full of gear was a far cry from today's pocket cell phones with their mobile Internet applications, but it was just the beginning of the computer network revolution.

Along the way toward building the Internet, the engineers, scientists, and computer experts working for ARPA invented modern computer graphics, e-mail, local area networking in the form of Ethernet, and of course the now-ubiquitous computer mouse. The vision articulated by Licklider at ARPA, Engelbart at SRI, and others who followed in their footsteps spawned not just the Internet but a major industry based on small computers so inexpensive and easy to use (compared to the old mainframes, anyway) that they were to find their way into every aspect of the lives of millions of ordinary people. Michael Dertouzos—a computer researcher who came to MIT as a student in 1964, became director of the university's Laboratory for Computer Science in 1974, and remained there through the 1990s—credited ARPA with funding "somewhere between a third and a half of the major innovations in computer science and technology."

Not a few of the computer scientists, programmers, and hardware engineers who created the ARPANET and laid the foundations of the Internet were deeply distrustful of the military during the unpopular Vietnam War that raged during the Internet's conception and birth, and it's to ARPA's credit that they came to devote their lives not just to this Pentagon-funded project, but to many other ARPA-funded experiments in information technology besides.

Some of these network pioneers attended meetings with high-ranking officers at the Pentagon wearing sneakers and even antiwar pins. One engineer joked to his colleagues that he thought he could pin his Ω, for "resistance," on the jacket of a general when he wasn't looking. These mavericks were not only tolerated but given complete autonomy by ARPA to pursue their research (and to dress) as they liked, in a policy that continues to this day.

At times the friction between military and academic interests have flared into public argument, as when students at the University of Illinois staged sit-ins in 1970 to protest the ARPA-funded Illiac IV computer project. With antiwar sentiment at a fever pitch, many students simply couldn't stand the idea of a mainframe computer dedicated mostly to military use being installed on their campus. The head of the project at the university succumbed to student pressure and bit the hand that fed him by publicly decrying his need to use military funds to accomplish his institution's own, more benign goals. "I know the military side of the Department of Defense (DOD)," he said in the student newspaper. "Some of them are dangerous fools. . . . If I could have gotten $30 million from the Red Cross I would not have messed with the DOD." As a result of the controversy, NASA took over the Illiac IV project and moved it to its Ames Research Center in California in 1975.

The ARPA director who took over in 1970, Stephen Lukasik, couldn't help but take such outbursts personally. "It is only in the Defense Department that you can get the freedom to do a lot of very important science," he pointed out in a 1975 interview. "I could understand the feelings that the Vietnam War engendered in people. . . . And yet I was always disappointed that people couldn't understand." Lukasik's message to researchers who took ARPA research funds and yet publicly slammed their benefactor: "This is *ARPA* and you know what we are doing and you know the way we play it."

DARPA is often still the only game in town funding advanced work in cutting-edge computer research. "The military has been much more accepting of new technology than many civilian industries,"

Mark Burstein, a computer scientist with BBN told me. He denied the existence of an ongoing culture clash between the Pentagon and the research community. "I do believe that all researchers should be aware of and make informed moral choices about the implications of their work on new technology, how it is paid for, and who has the rights to it. This is just as true when speaking of funding from commercial industry as it is from the government." Much of the work is still directly related to Licklider's vision for "man-machine symbiosis," and it still often proves equally beneficial to civilian and commercial applications as it does for the armed services.

To see what had become of some of the work that Licklider's vision had wrought, I decided to visit the world's second node on the Internet, SRI International, in Menlo Park, California. Here DARPA-funded researchers continue to engage in the advanced computer research it started in the 1960s, still breaking new ground not only in computer networking but also in artificial intelligence, speech recognition, and more.

SRI has worked hand in hand with DARPA as a contractor since before the early days of the ARPANET, occupying a sprawling campus in Menlo Park, California. The nonprofit research institute was born as the Stanford Research Institute in 1946, at Stanford University, in neighboring Palo Alto. In 1947 it moved to the site of the former U.S. Army Dibble Hospital, its current location, with a grand total of seven employees. It soon grew to encompass forty buildings on sixty-three acres at its Menlo Park headquarters and fourteen hundred employees at offices around the world, including in Washington, D.C., and Tokyo. The 1970 student protests pushed Stanford University administrators to sever ties with the institute because of its ARPA and other DOD research projects, and today it is known simply as SRI International.

The organization was founded as a nonprofit institute "to promote and foster the application of science in the development of commerce . . . the discovery and development of methods for the beneficial utilization of natural resources . . . and the improvement of the general standard of living and [is] dedicated to the peace and prosperity of mankind." A high-sounding ideal that the institute's CEO in the 2000s, Curt Carlson, puts more succinctly. "Our goal is to make this a better world," he says in an SRI promotional video, "one innovation at a time." In practice that means the institute engages in research for hire for various arms of the federal government and for private companies and other organizations around the world. SRI's first contract came from the U.S. Navy, and the Pentagon—including, after it was established, ARPA—has always been a major client of the institute. Today Department of Defense customers make up slightly more than half of the institute's client base. Although it's a nonprofit, SRI nevertheless functions in many ways as a for-profit business—attracting clients to fund its research and seeking ways to commercialize the innovations it develops as long as its clients allow it, with any money earned over and above that required to complete a given project (i.e., profit) getting plowed back into the institute's facilities and work to develop new projects that might attract funders.

As with DARPA itself, SRI's research takes in a breathtaking array of science and advanced technology in fields ranging from computer science and artificial intelligence to medicine, energy, and materials science. SRI also tackles problems in education and government policy, among other "soft" sciences. Along with the computer mouse, critical elements of the Internet, and many other now ubiquitous advances in information technology, researchers at SRI have also given the world the fax machine, high-definition TV, and telerobotic surgery among its fifty-thousand-and-counting research projects over its more than sixty-year history.

What I found on my visit in the summer of 2008 was something like a resort-style DARPA. The mild weather, the collegiate atmosphere, its location in a tree-lined residential neighborhood, the sandwich bar with grilled veggies and the salad bar at the cafeteria, the electric golf carts used to get around the campus, even the relaxed dress of the researchers and staff I met—shirts with open collars and no ties, often paired with jeans and even sneakers—all gave me the impression of a more laid-back, semitropical DARPA. In the spacious lobby and waiting room in A

Building, the institute's public face, a display along one wall showcased SRI's inventions and accomplishments over the years, including a pair of Emmys won for its work in high-definition TV. "SRI is the soul of Silicon Valley," reads a quote from *BusinessWeek* on the wall greeting visitors. "It was here before silicon chips, and its multidisciplinary focus was, and is, the foundation for the Valley's seething climate of innovation."

My chaperone for the visit walked me to my interviews with current researchers. On our way, we passed through the speech-technology department, with a wall of patents displayed on plaques. I also caught a glimpse of the Poulter Laboratory for "research on extreme deformation and failure of materials and structures," as the sign on the wall of the center's hallway termed it. "That's where they blow things up," my chaperone told me with a smile as we crossed the hallway, which was interspersed with emergency shower stations. I could only imagine the muffled "KRUMP!" of an explosion during the course of an ordinary workday at this place, followed by a brief cry and a puff of smoke issuing from a suddenly flung open door and a scientist with flaming hair running for a shower station to put himself out.

In a conference room I got a crash course—a sip from the SRI fire hose, as one of my interview subjects put it—on some of SRI's current work in information technology from a few of the institute's research managers. Up first, sixty-one-year-old senior scientist for the Speech Technology and Research (STAR) Laboratory, Jordan Cohen, filled me in on his group's work on systems that could accurately interpret speech and even translate it into other languages. DARPA's current Global Autonomous Language Exploitation, or GALE, program, one of the latest research efforts, aimed to demonstrate the ability to continuously monitor Arabic and Mandarin Chinese media feeds, automatically translate them into English, and deliver the information in them in response to queries made by a human operator.

Cohen's presentation had all the sex appeal of a university lecture, complete with diagrams on the conference room's white board. I wasn't able to get my hands on the actual system or even see it in action, and so the conversation stayed at an abstract level that was less than ideal for my understanding of the subject. This was a problem I was to encounter often during my research for this book, as researchers fought competing impulses to share their work with me and to obey the dictates of military research that required a lot of closed doors. Still, Cohen did his best to make me understand the work that had consumed his professional life since the 1970s, speaking slowly and carefully, and patiently answering my pleas to simplify his explanations so that a nonspecialist could understand it.

DARPA has been at the center of speech recognition research since the 1970s, along with the various intelligence (i.e., spy) agencies, funding independent research projects at universities across the country. "Of course it was invisible," Cohen told me, referring to the secret nature of these projects. "But there was a lot of money." From the beginning, the military's interest in speech-recognition technology had been twofold. "The government has two problems," said Cohen. "One is to keep our speech private from other people who might want to listen to it. The other is taking other people's speech that's obscured and making it intelligible. The difference between those two things is called the equities problem." Since so much of the work was about discovering ways to make one's own speech unintelligible to adversaries while at the same time cracking the deliberately unintelligible speech of those adversaries, the details of breakthroughs that the university-funded speech-recognition researchers made necessarily had to be kept under wraps. "You can't tell the 'ether' what you know," Cohen explained.

The technology had commercial appeal, of course, and so some of the research has gone on in the open, at places such as IBM's speech research group, in Yorktown Heights, New York, where Cohen went to work in the 1980s, after a stint with the National Security Agency. The computer giant's focus was on office automation, in this case dictation, and Cohen's task was to find new ways to improve the accuracy of speech recognition by machines. "There's always been a question about what it is you listen to when you listen to speech with a machine," Cohen told me of the field's central problem. "The standard thing is to do a Fourier transform and look at the power spectrum or some function of that." In other words, the most common method for getting a computer to recognize speech involves translating an analog voice signal coming from a microphone into a mathematical structure that the computer can actually work with. The work the computer does typically involves analyzing the frequency of the sound and looking for patterns that might represent phonemes, or particular speech sounds, and then comparing the sequences of phonemes to a vocabulary list to try to determine which words the speaker has used.

Cohen took a different tack in his work at IBM. He decided to examine the physics of what actually happened in the auditory system when listening to speech. Was it different from what the computers had been listening for in the standard speech-recognition systems? The answer, Cohen discovered, was yes. "It's a completely different function," he told me. In fact, his work for IBM in this area represented a breakthrough for the field of speech recognition. "I built the first successful auditory model," he told me matter-of-factly. Thanks to Cohen, the auditory model of speech recognition now forms a major part of cutting-edge speech-recognition systems. "I don't make any great claims to that," he was quick to point out to me. "It's just one of those things that happen in the history of science. Sometimes you sort of get things off in the right direction." It wasn't that auditory models hadn't been tried as the basis of speech-recognition systems. "I was just the first guy that made it work."

As early as the 1950s, researchers determined that people perceive sounds differently depending on context. For example, people perceive a sound that follows a louder sound as being softer than if it is preceded by a period of quiet. "Your nervous system has a sort of memory about what happened recently," said Cohen. "It's a big complex thing that's going on."

Applying this research, Cohen, and others who followed him, built a computer model of the way the human auditory system perceives speech sounds, used it to process incoming sounds, and then fed that data into a speech-recognition system, along with the output of the standard Fourier transform. The result was a more accurate interpretation of what was being said than what would have been arrived at using the standard signal processing technique alone. "It turns out," Cohen said of the combined approach used today, "you want a judicious mix of the two."

The art and science of speech recognition still has a long way to go. Recognizing speech in a range of conditions—crowded rooms, using different microphones, and among speakers with differing accents—presents a whole other realm of challenges for computers, and they are still only just being solved. Of course that doesn't stop DARPA, through contractors such as SRI, from continuing to work for the day when speech-recognition systems might be able to automate listening for and sorting through important information, and even translating that information from other languages.

Case in point, the GALE project that Cohen managed at SRI at the time of my visit. The program's mission was to incorporate four important functions into one system—something that had never before been attempted. The system was based on speech recognition, to be sure. But it would also have to be able to translate the speech it heard in one language into another language. Once translated, the information would have to be annotated to make it easily searchable by a human operator. Finally, the system would have to deliver the information to the user. The result would be a powerful intelligence-gathering tool. To make the system effective, said Cohen, "You need to find geopolitical entities; you need to find times and dates." In addition, "There are a few verbs that you know are going to be interesting, so you try to find arguments for things like *attack* and *arrest*—the 'who,' 'what,' 'where'—and you try to annotate that in the database, so when people go to ask questions, the database actually has some semantic understanding as well as just the text."

Three years into the five-year program, SRI's version of the system (DARPA had BBN and IBM working on competing versions) was engaged in parsing, translating, and indexing Arabic blogs. The country's adversaries, said Cohen, "the people that are sort of on the other side from us, tend to want to make a big show of things, and so they both tell you what they're going to do, and after it's done they take pictures and tell you all about it. So there's a lot of data around." The goal was to mine that data for intelligence that could be acted on. The state of the art had a long way to go, however. "We're nowhere near being able to handle that data."

Not least of the project's challenges was the translation component. Even assuming that the experimental system could accurately recognize Arabic speech across a wide range of recording environments and the speech patterns of individual speakers, there still remained the very difficult task of translating that speech into a form readable to an English speaker—and just as significantly, measuring the success of the translation with a view toward improving future iterations of the system. DARPA director Tether had set as a goal that the finished system should have a translation accuracy rate of 95 percent. Trouble was, how do you measure accuracy? As Cohen pointed out to me, "There are so many ways to say things, all of which are okay. You could change around the clauses, you could use synonyms for things, you could use expressions for single words—all kinds of stuff you can use to do a translation. In order to measure whether it's right or not, what you want to know is: is the semantics of the thing that you produce [with a computer] the same as the semantics of the reference? We don't know how to automate that, and we're still searching for that."

The current measuring scheme involved a laborious process of hiring a cadre of four human translators to translate the same block of text without comparing notes. Then a human editor compared the human-translated versions with the same chunk of text as translated by a computer, and attempted to write his own version of the text that somehow bridged all of the translations. "He writes a sentence that is semantically the same as the four human outputs," Cohen explained, "but it's as much like the machine output as he can make it." The last step of the scoring process required comparing the output of the human editor with the output of the machine to see how much they differed.

Besides being extremely time consuming—it took forty editors a month and a half to score a test—this inherently subjective process wasn't even particularly accurate, according to Cohen. "As a sanity check," he said, "we just took a test, the kind of test we do every year, and farmed it out to a translation house and asked them to deliver us human translations." Applying the same test that DARPA used to track the progress of the computer translations, the GALE team found that the *human* translators flunked the test. "They're not at 95 percent," Cohen said. "So humans can't do it." Clearly the way in which the performance of the computers was being measured was flawed. "There's going to have to be an adjustment." In the meantime, the existing test was the best DARPA had. And the systems in development *were* undeniably improving. "We're sort of grabbing on to the very tail of making this sensible," Cohen told me.

Up next in the conference room, to take their turn blasting me with the SRI fire hose, were Artificial Intelligence Center director Ray Perrault and Information and Computing Sciences Division manager Bill Mark. Together the two men, both in their late fifties, headed SRI's Cognitive Assistant that Learns and Organizes, or CALO, project, which was in turn part of DARPA's Personal Assistant that Learns, or PAL program. The aim of both projects was nothing less than the creation of machines intelligent enough to anticipate the needs of their users and perform routine tasks for them automatically. I'd seen a DARPA video promo at DARPATech that showed a three-dimensional computer projection of an artificial solider reporting to a superior human officer on the status of a military operation and receiving orders.

The reality of the actual system, now in its fifth and final year of development, was much more mundane. When I asked PAL's program manager at DARPA, David Gunning, whether it was a coincidence that the name of his program rhymed with HAL, the name of the murderous, but highly intelligent, computer in the 1968 film *2001: A Space Odyssey*, he laughed. "You know," he said, "we're so inundated with science fiction movies and TV shows that make it seem like that's just around the corner . . . and you always see . . . articles about 'are we afraid that computers are going to become self-aware?' I think someday we may need to worry about that, but I don't think we're close to that yet."

The much more modest goal for PAL was to allow computers to do some of their own programming by having them learn new tasks in ways that more closely approximated the way people did so—through observation and natural language instruction rather than through the laborious process of hand-coding each and every instruction.

Gunning was in his second stint as a DARPA program manager. The first had been in the mid-1990s, when he and his colleagues had approached this same problem by trying to simply pack computers with as much information as possible in an effort to achieve a critical mass of understanding. "The great hope for creating intelligent computers," Gunning explained to me, "was to create a knowledge base where you go in and have humans encode knowledge—in mathematical logic—in a computer database, and hope that if you added enough of this information to a knowledge base it could do more and more intelligent reasoning." The major drawback to that approach was the sheer volume of the information that had to be entered into those databases, and by some very expensive labor, too, "by some very high-powered engineers. Because it's a very subtle, complex job to take what seems like simple, everyday knowledge to a human, but then express that in mathematical logic in a way that the computer will then do correct reasoning."

Much of Gunning's efforts at the time involved developing more efficient ways to load up those databases—first by streamlining the process for computer programmers, and then by developing software tools to allow people who actually had the specialized information, such as military tactics that needed to be entered into the databases, to do it themselves. "But it still was just too difficult to enter this knowledge into the knowledge base," Gunning told me. Not all was lost, however. At the tail end of his first tour as DARPA program manager, Gunning put together a slick interface for a knowledge-based system, a visualization front end, as he termed it to me, with the goal of showing potential military users the advantages of knowledge-based systems. "The visualization stuff we created was very popular," he said, "and ended up being a success, but the knowledge-based stuff was still too difficult." Gunning returned for a second tour as program manager in 2003. By then the visualization program he had helped to create was in use by the army as the Command Post of the Future, or CPOF, and while it didn't fulfill Gunning's goal for reasoning computer systems, it did represent a breakthrough in its own right, in part by realizing a part of the dream of Doug Engelbart—he of the Mother of All Demos.

Perhaps because of his work on time sharing and networked computers, Engelbart had always imagined advances in interactive computing—the mouse, displays, word processing, and all the rest—as simply the means to a greater end. One part of that end was, as he put it, to "augment" the human intellect by combining the best features of human and machine. By "enhancing the human intellect," Engelbart didn't mean just that computers would become better at serving the needs of people, but that people would also stretch their mental capacities to meet machines somewhere in the middle. Hence, during the Mother of All Demos, Engelbart demonstrated not just a mouse and a keyboard as input devices, but also a five-button chording keyboard that allowed its user to type all the letters of the alphabet one-handedly by pressing combinations of keys at the same time. To this day, Engelbart swears by this novel keyset as a way to more efficiently enter information into a computer. He imagined humans training themselves in all kinds of ways, *evolving*, to better interact with their machine counterparts while at the same time expanding their innate capacity to reason.

Needless to say, the evolution of computers has far outstripped that of their masters. Most humans haven't adapted much more than to learn to use a mouse for interacting with their computers. As a former colleague of Engelbart's, Don Nielson, put it in his 2006 book, *A Heritage of Innovation: SRI's First 50 Years*, "The direction that personal computing is taking is to help make us more versatile in what we can do easily, but nowhere is there a staircase of computing sophistication that takes a user to a higher effective plane of thought—which is, after all, what augmenting intelligence really means."

But another part of Engelbart's far-reaching vision, the one advanced by DARPA program manager David Gunning's visualization program, is now being realized. Besides augmenting the human intellect, Engelbart also wanted to create the tools by which networked groups of *people* could more easily collaborate. They would do this by forming networks of communities that would share resources and ideas to accomplish more, and more efficiently, than individuals could do working by themselves. Said Nielson in *A Heritage of Innovation*, Engelbart wanted to enable "groups, collaborating through networked computers, across space, to accomplish what would have been impossible before."

DARPA's Command Post of the Future did just that by allowing army commanders, logistics planners, and intelligence specialists to share information related to a planned mission and collaborate on the best way to deploy their resources. The system graphically represented the mission on a map of the area in which the mission was to take place. The collaborators could click, drag, and drop text and images to share information and update the map that they all saw before them. For example, Gunning explained to me, "an opps guy develops a plan, and now he wants to pass it off to the logistics guy to support it. He just kind of drags that plan over to the logistics guy's map, and, boom, it's all there."

The key feature of the system was that each member of the commander's team kept his or her own view of the situation, while the commander kept a top-level view. "But more important," said Gunning, "the commander can look at each of his staff's visualizations and get a better understanding of what they were thinking, actually, than he did if they were all in the same room trying to carry on a conversation." It was exactly the kind of system Engelbart imagined when he and his team created the mouse and all the other interactive computing tools.

And now Gunning, in his second tour at DARPA, wanted to upgrade CPOF with the capabilities he had planned for it from the beginning:

the ability to reason about the work the humans were doing on it—in effect, to become another member of the commander's team. Shortly before Gunning came back to DARPA, director Tether had hired a new IPTO director named Ron Brachman, whose main mission, as Gunning put it to me, "was to create what he called cognitive software, cognitive systems, which were not only systems that were intelligent, but systems that could learn. So that is the key to this knowledge acquisition problem, right? If you could get computers to learn on their own, just bumble around like people do, eventually you figure out what's going on." It was Brachman who brought Gunning back to help realize that agenda. "To me," Gunning told me, "this was exactly the right solution to this problem." If Gunning and Brachman were successful, they'd eliminate the need for programmers to spend hours and hours hand-coding information into databases to make computers smarter. Their computers would gather the information on their own, effectively programming themselves. To realize this vision, Brachman and Gunning put together the PAL project.

Gunning said, "Imagine you had your own assistant inside your computer. It would watch what you do and learn: What topics are you interested in? Which e-mails do you respond to? How do you schedule meetings? How do you like to prepare your briefings? What tasks do you do? It could watch as you planned a trip. Watch as you did a literature search. And then gradually learn what your preferences are, how you do things. And then, gradually, it would be able to execute more of those tasks for you. Eventually it's going to be able to learn more and more of your tasks. It will be building up its own knowledge base about you and just about how things are done in the world."

The implications were profound. Such self-programming thinking machines could change the way we interacted with computers as thoroughly as Engelbart's interactive computing tools or the Internet had done. DARPA director Tether put it to me this way: "We are going to have computers that will learn *you* as opposed to you having to learn them. What this will mean is that your computer will be with you from the moment [you're born and as] you grow up, and it will, over time, get to know you, know what your habits are, know what you want to see, know how you take in information, know when you want information. How that's going to impact the world, I can't imagine. But I do know that it will be a big game changer."

Doing the work of creating that game changer was a team of twenty universities and a few companies and research institutions, including SRI, all led by Bill Mark and Ray Perrault, who had taken their seats in the SRI conference room with me as I was finishing up with Jordan Cohen. Unlike Cohen, these two had brought demos with them, which put me a step beyond the mostly theoretical level of understanding I had with Cohen's project. But here again, I didn't have a chance to use the system. The demos were mere static videos of the systems at work, accompanied by voice-over narration.

Since the software they were working on was designed to run on ordinary Windows PCs, I asked the men the obvious question: How about giving me a copy to take home and bang around on to give me a sense of how it worked? "We can certainly show you a demonstration of it," Mark replied. "We can't give you a copy of it." Naturally, I pressed. "We're not being hesitant or secretive," returned Mark a little more firmly. "There is a release process for this information, and we will give you things that have gone through the release process." In other words, they *were* being hesitant and secretive. They had to be.

Of course I was disappointed, especially since program manager David Gunning at DARPA had told me that the project had created an installation disk of a slimmed-down version of the software called CALO Express, suitable for popping into any computer's DVD drive and running it. "Now it's available on DVD," Gunning had told me. "I'll put you in contact with SRI and they can, you know, you can work with them, and they can install it on your computer." Gunning must have known at the time that that wasn't going to happen even as he stumbled over that sentence. And yet . . . why create an installation DVD in the first place?

"In one of our reviews there," Gunning explained, "Dr. Tether said, 'Hey, this stuff is so great, why don't you guys use it yourselves at IPTO?'" The disks had been created for people at DARPA itself to use. Easier said than done. It took Bill Mark and Ray Perrault and their team at SRI a solid year to create CALO Express, and then for a whole year after that the team struggled, as Perrault put it to me in our meeting, "to actually get it past the guardians of the DARPA IT system. Finally, actually, we gave up, and we found other users whose organizations were more welcoming." One such organization was the U.S. Navy.

Seems the IT Department at DARPA, true to IT departments everywhere, was reluctant to install unvetted software on their systems. More than that, it apparently had no intention of unleashing some experimental self-teaching, self-programming software package on DARPA's computer infrastructure. Come to think of it, maybe I didn't want such a thing running around loose on my system, either. Whom would I call for tech support? Still, even an in-house live demo at SRI would have been nice for my understanding of how the package worked, but no such luck.

Perrault and Mark hoped that the demo video Perrault showed me on his MacBook Pro would give me a general sense of what the systems they were building could do. But I was further handicapped by the fact that the full-up version of CALO was designed for the Command Post of the Future, itself an arcane piece of software I had no familiarity with. All I knew was that the CPOF ran on top of Windows, made use of a three-screen display, and facilitated military mission planning with a map-based interface. The video showed CPOF in action with CALO running on top of it, and I had a hard time making much sense of what was going on.

"As Major Davis selects Execute in step one," the narrator for the video intoned, "the mission statement appears. When the G-3 is satisfied, he clicks Resume Execution to move to step two: Commander's Intent. When Major Davis developed the cordon-and-search procedure, he was able to import and use live data from the CPOF repository. When the repository data is updated, that data is automatically revised in the learned procedure. During the briefing, the commander modifies his intent and adds a key task to conduct a meeting with the local leaders to coordinate stability operations. Major Davis selects the PAL Watch function and makes the changes. This ensures that any changes or updates made are learned and incorporated by the procedure." Huh?

Best I could figure, the commander had launched a pre-learned procedure, told the copy of CALO (i.e., PAL) running on the system to watch what he was about to do, and then modified the procedure so that the modification would become part of the original procedure. The system also drew from continually updated information contributed by other users and automatically added it to the procedure the commander was working on—in this case a cordon-and-search procedure. I thought I was beginning to understand. The system provided a shortcut to constructing and saving complex procedures, and those procedures could adapt themselves to changing conditions—all without the kind of hand-holding and expertise that more traditional programming required. "A PAL-enhanced CPOF will be able to learn and assist users in an environment where time is lives," the narrator concluded.

Flexibility was the key to CALO's success, an ability to adapt to changing conditions and to its users—traits that computers weren't best known for. I was familiar with computer macros, that is, recorded sequences of commands or keystrokes that someone using a computer plays back when needed, and I asked Perrault how CALO differed. "It has to draw inferences between the inputs and outputs of various steps and link them in ways that macros typically don't have to do," he told me. "The outputs from one feed in as the inputs to the other." Uh-oh, I was getting lost again.

But the two computer scientists helped me out with an example of the system at work that I could better understand than a military application I had never used: Web searches. The next video Perrault played on his laptop had a user training CALO to search for bed-and-breakfasts within two miles of a given address. "Click the Search button," the human said out loud to the computer.

"All right," said the computer in a synthesized woman's voice that sounded a bit like a German tourist reading from a phrase book.

"Here is the list of results," said the human. He spoke so slowly and carefully himself that I had to ask Mark and Perrault to make sure that this was indeed the human speaking.

"Can you check that I understand the data?" the computer asked.

"This is the name of the bed-and-breakfast," said the human, clicking it.

"Okay."

"And this is the distance."

"All right."

"Click Next to get more results."

"When should I stop searching?"

"Get three pages," said the human.

After a pause: "I am finished," the computer said.

"Save bed-and-breakfasts closer than two miles."

"I don't understand," said the computer after a pause during which it returned a speech-recognition error, according to Perrault.

"Save bed-and-breakfasts closer than two miles," the human repeated with saintly patience.

"Okay."

I was still having a tough time understanding how a system like this would help me to work better or faster. Listening to the demo was a bit like hearing a patient parent teaching a toddler how to use a computer—something I could definitely relate to ("No, no, click the *other* mouse button"). And generally, I get more done at work when I leave the kids at home. The SRI team had anticipated this kind of reaction, so they had built in certain implicit learning faculties, along with the explicit ones. "We don't want to annoy the user by constantly forcing them to give feedback," Perrault explained to me. He gave me the example of a desktop search function. When the computer assistant returned a group of results, it would watch which files listed in the results you actually clicked on to learn the parameters of useful search results. "So, if you get a result," said Perrault, "and you open that file, it says, 'Aha, that must have been a good result.'" It also would learn from the files you *didn't* open, to know which kinds of files to exclude from future searches. In this way the computer's search results would continually improve the more you used it, without having to bug you with questions and requests for confirmation. "It doesn't learn these things with one hundred percent accuracy," said Perrault. "It just says, 'you know, I think this is right.' Somewhat like people learning."

All this seemed like pretty mundane stuff to me, even if it did work as advertised. Was the state of the art really this far away from HAL? "DARPA, in conceiving the PAL program," Bill Mark told me, "is trying to revolution-ize software. I mean, it really is that big." Maybe I'd have to be a program-mer myself to understand how SRI's work on CALO could progress toward Gunning's ultimate vision. Of course, most people hadn't understood the early work of Doug Engelbart, either—not until he delivered the Mother of All Demos. "The vocabulary doesn't seem to exist with which to commu-nicate briefly the subject range involved in this multidisciplinary study," Engelbart had written in an early request for funding. Perhaps that was so here as well—that CALO would have to progress further toward the realm of practicality for nonspecialists like me to see its revolutionary aspects.

Obviously proud of their achievements thus far, Mark and Perrault did their best to show me the project that Perrault described to me as the highlight of his career through their eyes—without actually letting me see the fruits of their labor. In they end, to their credit, they succeeded, at least in part.

The killer app for the project, at least initially, might turn out to be something like the part of CALO the two computer scientists demoed to me next: a meeting assistant. The narrator for this video described a system that would help meeting participants pull together relevant files before the meeting, share those files with their colleagues, and then—and here was the kicker—record the entire meeting while making a transcript of it on the fly. Afterward, the system would tease out the main topics from the transcript, identify action items for each participant, and highlight the decisions the participants had reached. All of this would be packed up into e-mails for each participant, thus sparing everyone the tedium of actually having to read the transcript. "Nobody wants to look at a meeting transcript," Mark explained. "I've done it. They're extremely dull and annoying. However, if the system can look over that transcript and get an idea of the topic that was being discussed, or look at action items or decisions, it can then provide a meeting summary which is much nicer to look at."

The transcript wouldn't be perfect. "It's not going to be understanding every word," Mark told me. "It would make some mistakes." In fact, the transcript wouldn't be so much a transcript, from what I could gather from Mark. "Let's call it the gist," he said. But since each person in the meeting would be wearing his or her own microphone, the system would know who was speaking when, greatly aiding the process. When you got back to your desk, you could open up the transcript and meeting summary the system had e-mailed you, and anytime you got to a section that didn't make sense, you could click the section to hear that portion of the meeting played back.

Now *that's* something I could use. I had to transcribe my audio recordings of my meetings with Cohen, Mark, and Perrault at SRI myself, a process that took many times the duration of the actual interviews. It would have been very nice indeed to have been able to glance over even an approximation of what had been said and then jump to particularly interesting quotes for verbatim playback. Probably I wouldn't have any problem inducing Mark and Perrault to wear microphones to our meeting, either. However, it was unlikely I'd be able to get my research subjects at other locations to put them on, even if I'd wanted to lug them around with a laptop to each interview. On second thought, maybe I'd be better off learning shorthand.

If Cohen, Mark, and Perrault's presentations were necessarily less than ideal for my appreciation of their work, my visit to SRI's Trauma Pod lab more than made up for it.

THE ROBOT WILL SEE YOU NOW

THE PATIENT WAS IN BAD SHAPE. He lay on the operating table with the lower left quadrant of his abdomen open to the air, exposing the redness inside. The four robot arms of a da Vinci surgical robot hung poised over the wound like the forelegs and mouthparts of a giant praying mantis. But the machine was still and silent, as was the robotic scrub nurse on the patient's other side. For that matter, so was the patient himself. In fact, he'd been lying there for the last year and a half. Not that he minded particularly—he was made of rubber.

I stood in the operating room with forty-seven-year-old mechanical engineer Thomas Low, director of Medical Systems and Devices at SRI International. One of his teams of engineers—led by principal investigator Pablo Garcia—had created this, the first fully automated surgical suite, and they were maintaining it in stasis while they awaited the funding they needed to resume the project where it had left off following a successful demonstration for DARPA back in early 2007.

The wait wasn't exactly DARPA's fault. The agency had pushed the project, called the Trauma Pod, at typical breakneck speed. But given director Tony Tether's focus on finding homes for projects among the armed services, the program had to wait for its new home, the army, to clear all the necessary approvals before continuing. And so the robotic surgical suite sat, ready for demonstrations at a moment's notice, ready for modifications in its next planned phase.

The Trauma Pod, at the time of my visit to SRI in the summer of 2008, was a project in transition. Based on SRI's work with surgical robots in the mid-1980s, the Trauma Pod became an SRI-led DARPA program in the 2000s. In 2007 the Trauma Pod project completed phase one, demonstrating the basic technologies needed to care for critically wounded soldiers on a battlefield in the absence of surgeons or even a field hospital. DARPA program manager and surgeon Richard Satava envisioned the Trauma Pod as an automated self-contained emergency suite. Soldiers in the field would load a wounded comrade into the system, jump out, and close the hatch. Inside, robots would scan the wounded soldier with a magnetic resonance imaging system to diagnose him, and perform one of several life-saving procedures to stabilize him for the ride, on either a truck or a helicopter, to a fully equipped hospital.

"Thirty minutes to an hour," SRI's Low explained to me was the estimated time of arrival for a Trauma Pod–encapsulated soldier from the battlefield to the hospital. Any longer than that and all bets were off on a seriously wounded soldier's chances for survival. But the Trauma Pod could mean the crucial difference between life and death for soldiers with wounds that surely would kill them without the pod keeping them alive for those crucial minutes.

Low had been working at SRI since earning his undergraduate degree from the University of California at Berkeley in the early 1980s. He got his graduate degree from Stanford, practically next door to SRI. "I was always interested in medicine," he told me. "You know, *The Six Million Dollar Man*. I grew up with that," he said, referring to the popular TV show of the 1970s featuring an injured air force officer and astronaut restored to better-than-human health with robotic parts following an airplane crash. "Biomedical engineering seemed a real cool thing," Low recalled. And at least partly due to that inspiration, he told me, "Robotics was the thing that I was really interested in as an undergrad." When I asked Low if he read science fiction as an adult, he replied, "A little. When I'm not busy creating it, thank you!" Somehow along the way he found the time to become the third-ranked aerobatic hang-gliding pilot in the world and to build his own Lancair airplane. His proudest accomplishment, however, is raising his three kids.

Low is personable and genuine, but he talks rapidly, like so many other researchers and managers I met during my travels, as though he has no time to lose in explaining the Trauma Pod system and how it works. It's almost as if, by explaining it more quickly, he could just as quickly put the first phase of the project to rest and move on to phase two.

SRI researchers invented the field of modern surgical robotics. Low came on board in 1984, and he was part of the action from the beginning. In those early days, the mechanical engineering department of which he is a part was quite distinct from the medical group. "Back then we were working with industrial-type robots," Low told me. "Really, for the longest time after joining SRI in eighty-four, I was doing a lot of automation work, mostly post office kind of things, handling mail. It wasn't robotics like this." But some in the medical group, particularly a researcher named Phil Green, began to see the possibilities in robotic systems applied to medical applications. And to implement those ideas, Green and his crew made forays into the mechanical engineering department to borrow the expertise of their nonmedical brethren. "So we became involved in the development of the systems," Low told me.

Low and his crew and Green's group merged into the Medical Systems and Devices Department, with Green as the head—the position in which Low later succeeded him. It was Green and his original medical group who came up with the idea of pairing a so-called master robotic system with a mirroring system known as a slave to perform medical procedures by giving a surgeon the ability to project his or her abilities into a remote location. These were not to be autonomous robots, robots that could act by themselves without intervention, but robots designed to reproduce motions initiated by humans.

That in itself wasn't a new idea. Telemanipulators, or waldos (the name comes from a 1942 short story by science fiction writer Robert Heinlein) had been used as early as 1945, when a company known as Central Research Laboratories built a system to let an operator handle dangerously radioactive materials without touching them himself. Green's innovation was to transfer the *vision* of the operator along with his physical movements to the remote location, and that made all the difference in allowing such systems to be used not just for gross movements such as shoving big blocks of material around, but also for the delicate movements, requiring a lot of hand-eye coordination, needed for surgery.

In the medical robotics lab, a neat, largish room—especially compared with the cramped quarters of APL's Revolutionizing Prosthetics lab—with windows lining one wall and a couple of engineers working code on computer workstations, Low showed me an early version of the system that revolutionized a class of minimally invasive surgery. It was a tall green metal-and-plastic box with what looked like binoculars poised over a tabletop. The device was called the M4, after the number of degrees of freedom, or independent motions, it was capable of. (By contrast, the human hand and arm have more than twenty degrees of freedom.) Seated at the machine, an operator could peer through the binoculars and seem to be looking down at his own hands placed on the master controls just below. But in fact, he would be looking at twin video displays, one for each eye, receiving feeds from a pair of cameras on the slave manipulator across the room or miles away. The three-dimensional view would show the robot slave's graspers and other tools with which it might be equipped. When the operator moved the controls, the slave's manipulator arms would make the corresponding movements. After a few seconds, Low told me, the operator would forget that he wasn't watching his own hands.

To complete the illusion for the operator of being physically present in the remote location, Green and his team added a haptic sense to the system, or sense of touch. Motors on the master system would push back on an operator's hands when the *slave* system pressed into an object. Finally, said Low, "we re-created the audio field." Microphones in the remote location picked up the sounds of the robot arms and manipulators working with their environment and transmitted them to the operator. All in the service of creating in the operator a suspension of disbelief. And if a system can do that, Low told me, "You suddenly start thinking you're there, and you start performing as if you're there."

From the beginning, Green and Low and their colleagues envisioned their system as a true telerobotic system. Actually more than that, a tele-*presence* system, one that would project not just a user's abilities to another location, but his sense of presence as well. They thought the system would prove its value in rural settings far removed from medical specialists in major population centers. Unfortunately, when Green submitted a proposal to fund ongoing work on the concept in 1989, the National Institutes of Health, or NIH, didn't agree. So the team built a limited-capability demonstration system with the funds they were able to cobble together from within SRI itself. The system had controls and manipulators, or effectors, as the group called them, for just one hand. But it was enough to demonstrate the unique capabilities of the system. As early as 1987 the work had piqued the interest of army surgeon Colonel Richard Satava when he and Green met at a medical conference. Satava was then chief of general surgery at Fort Ord, near Monterrey, California, and he was to become instrumental to SRI's continuing work on robotic surgery in general, and its later work with DARPA in particular.

Satava had been a NASA astronaut candidate, and remained keenly interested in the use of virtual reality for training astronauts. He joined the SRI team as a consultant while continuing to work with NASA, and that allowed him to make a crucial mental leap. "I made the connection," Satava told me later, "that if you sat down at a workstation and you used the robot, you could actually import the patient's image [into the computer], and . . . do virtual reality surgery and practice the operation without making mistakes on the patient. Then, when you had the right operation you [could] just flick the switch and do it on the actual patient."

Information technology was revolutionizing fields from writing to manufacturing, reasoned Satava. Why not medicine? The knowledge required to operate on a patient—the patient's unique anatomy and the nature of his or her injury, and the procedures used to patch him up—

were just information, after all. If that information could be translated into code that a machine could understand, there was no reason robots couldn't be built to perform surgical procedures, thought Satava.

This would open up all kinds of opportunities to bring the operating room to the patient, rather than the other way around. Surgery in outer space, in remote, sparsely populated areas, even in war zones—all would become possible. The immediate application Satava wanted to realize was for surgery on the battlefield, and he secured limited Department of Defense funding to supplement SRI's internal financing.

"The other thing I brought to it," Satava explained to me of the early telerobotic surgical system at SRI, "was the critical need for this to be a completely intuitive interface." The Apple company had by then made a name for itself as a maker of easy-to-use personal computers, and Satava figured that was a good standard to apply to SRI's robot surgical system, "so that when someone looked at it, they would know what to do . . . and there wouldn't need to be any instructions."

The team at SRI, now aided by Satava as a consulting surgeon, completed its one-armed system and, in 1991, again went after NIH funding. After the NIH shot down that proposal, too, Green began to feel a little desperate. At a cost of $250,000 and counting for his project, he knew he was about at the limit of what he could pull out of SRI. He simply had to get outside funding or the project would die on the vine. Left with few other options, he made a direct personal appeal to the head of the NIH program office through which he wanted to get funding. The gamble paid off. Impressed by a video of the SRI system in action—not actually doing surgery, but being used to peel a grape and to paint a flower, which was proudly signed by SRI researcher Lynn Mortensen as "the world's first Telepresence Artist"—the program office director set the SRI team up with a three-year contract in 1991 to continue to refine the demonstration system, which came to be known as the Green Telepresence Surgical System.

More demonstrations, more videos, followed in 1991 and 1992. "I began going around to conferences showing it," Satava told me of the system. At one of the conferences he met Colonel James Peake, who was then the army's deputy surgeon general. Peake saw the potential of the Green Telepresence Surgical System immediately. He took the robotic surgery idea to DARPA and advanced Satava as the program manager. The agency hired Satava in 1992.

As he told me later, when he walked in the door Satava had no better idea of what DARPA was than the average citizen—that is to say, next to nothing—but he was surprised to find himself the only MD at such a well-funded scientific research agency. A program to train dolphins to use their natural sonar to find underwater mines "was literally the only life sciences program," Satava said. He went to work immediately to rectify that situation.

The first order of business was to create a DARPA program for funding SRI's work in robotic surgery. Satava called the program MedFast, and by 1996 the system was in good enough shape for Satava to run a demo of it at the SRI campus for the TV program *Scientific American Frontiers*, hosted by actor Alan Alda, who had made his mark playing surgeon Hawkeye Pierce on the 1970s TV series *M*A*S*H*. With the slave system set up in an armored personnel carrier and the master connected to it by a cable in a tent a few hundred feet away, Alda was able to give the system a test drive on a mannequin stuffed with pig intestines. "The most remarkable thing about the experience," Alda raved afterward, "was that the computer interface between the instruments and my hands allowed me to *feel* what I was doing."

For Satava that was just the beginning. "What became apparent after I started working at DARPA," he told me, "is many of the things I had read as a teenager in science fiction in the fifties we could now do." Things such as building Dr. McCoy's tricorder from the *Star Trek* TV series, in the form of a handheld ultrasound machine dubbed the SonoSite and built by an ATL Ultrasound spinoff company of the same name. "I searched my memory for those things I read about in science fiction and began building them," Satava told me, "because the technology was now available."

In fact, Satava wasn't content with just ransacking old-time science

fiction for ideas. He went right to the source. "I put on a couple of conferences while I was at DARPA, with science fiction writers," he explained.
"From time to time I would pick their brains. You know, have a conference, ask them to come and tell me what they thought about the future
of battlefield medicine or the future of health care in general or just the
future directions of various areas." Satava pulled together these ideas into
a collection of projects he called the Advanced Biomedical Technology
Program, with what became the Trauma Pod as the centerpiece. There
was a smart T-shirt designed to monitor the vital signs of a wounded
soldier. There was the LSTAT (Life Support for Trauma and Transport),
designed to function as a kind of interface between a patient and surgical
robots, and resembling nothing so much as the diagnostic sick bay beds
from *Star Trek*.

One innovation that entered into general use was a DARPA-funded
digital X-ray system that started as a project at General Electric called,
grandly enough, the Apollo Project. Starting in about 1993, Satava
funded competing approaches at GE and Xerox to create a diagnostic
system capable of transmitting digital X-ray images from the battlefield
to a field hospital. "Up until then," Satava told me, "you would take a film
and then you would run it through a digitizer."

GE won the runoff between the two companies to demonstrate a
working system in 1995. DARPA's financial contribution amounted to
about $10 million by 2001. "We did it," said Satava, "because it sup-
ported the Trauma Pod program. We wanted a scanner in there." By the
late 1990s, digital X-ray technology was well on its way toward replacing
conventional X-ray film at major hospitals. GE didn't limit the technology to general-purpose diagnostics, as in conventional X-ray machines.
In 2000, the company unveiled what it billed as "the world's first digital
mammography system." To be sure, GE had been at work on this technology before Satava came along, but the money from DARPA came at a
crucial juncture for the company, shaving a full one to two years off the
development process.

Satava saw the Trauma Pod as the core of his Advanced Biomedical Technology Program, and as with many of his other projects, he found inspiration for it in science fiction. "It was taken from Robert Heinlein's [1959] book *Starship Troopers*." In a 2005 paper, Satava explained further that in the novel, "the casualty was placed inside this cocoon or pod, which was imagined to be a combination intensive care unit (ICU) and operating room (OR), capable of completely rescuing and, if necessary, operating upon a wounded soldier while being returned safely to the spaceship." Satava may be mixing up his science fiction references. The novel actually has soldiers stripping wounded soldiers of their battle armor (which displays the wearers' vital signs to their comrades, à la Satava's smart T-shirt) before tossing them naked into a "retrieval boat" to join their uninjured comrades for the ride back to their starship, where medical care presumably awaited them. I say "presumably" because the book doesn't actually address care for wounded soldiers. But where Heinlein was silent on the subject of future medicine, another science fiction writer, Larry Niven, was eloquent. Satava's Trauma Pod in fact bears a striking similarity to the autodoc, which appears in many of Niven's short stories and novels, including in this passage from *World of Ptavvs*, published in 1966:

> *Luke pulled himself out of the tank and read the itemized bill. It was a long one. The 'doc had hooked by induction into his spine and done deep knee bends to build up muscle tone; recharged the tiny battery in his heart; and added hormones and more esoteric substances to his bloodstream.*

"The sound bite," Satava said, "was 'the golden hour becomes the golden minute.'" Helicopters had revolutionized medical treatment for soldiers by airlifting them right from the battlefield to field hospitals, giving anyone who could live through the "golden hour" it took to get them there a fighting chance for long-term survival. What Satava was telling me was that the goal of the Trauma Pod program was to cut that golden hour down to a "golden minute," where all a wounded soldier would have to do to have a chance to survive his injuries would be to hang on for just one minute after being wounded. An ambitious goal indeed, one that would be accomplished by effectively bringing the hospital to the wounded.

Satava left DARPA in 1999, during the tenure of director Frank Fernandez, and came back in 2002, the year after Tony Tether took over as director. In the interim, the Trauma Pod project lay fallow at DARPA, but the robotic surgery team at SRI stayed busy trying to commercialize their invention.

One of SRI's missions is to spin off successful technologies developed in-house into successful independent companies, and the managers at SRI's Medical Systems and Devices Department looked for an opportunity to do so with the Green Telepresence Surgical System. "We pitched it to every VC [venture capitalist] we could think of," SRI's Tom Low told me. "And we got nowhere." The medical community just didn't perceive the need for such a contraption.

The institute has maintained its position at the forefront of technological development since the 1940s in part by marrying cutting edge science and engineering with good business sense. Sometimes the technologists need a leg up from the business types, and that's where someone such as physician Fred Moll came in. Brought in to realize the business potential of some of SRI's medical technologies, including the telerobotic surgical system, Moll quickly realized that the system's commercial value lay not in remote manipulation, but in allowing finer and more intuitive control over surgical procedures taking place in the same room. As Low later recalled, "He said, 'You know, there is this new thing called minimally invasive surgery.'"

Minimally invasive surgery defines a broad category of surgery of which laparoscopic surgery is a major part. Laparoscopic surgery requires the surgeon to make small openings, or ports, in a patient's belly, through which he first inflates the abdominal cavity with carbon dioxide to get a decent amount of room in which to work, and then extends a miniature camera and specially designed small-bore surgical instruments through the ports to reach the organs to be operated on. The advantages of two or three quarter-inch ports versus a two- to four-inch or even longer incision are obvious—the less the patient has to be cut open, the shorter and less painful his or her recovery.

That benefit is certainly obvious to patients and hospitals, both of which have every incentive to minimize hospital stays as much as possible. And surgeons going into the new specialty of laparoscopic surgery in the mid- to late 1980s were only too happy to offer the procedure, regardless of whether it was more effective at treating a particular problem. Not everyone was convinced of the benefits of minimally invasive surgery, however.

Andrew Kagan, a vascular surgeon who practiced at Hackensack Hospital in New Jersey, considered extending his practice into laparoscopic surgery in the mid-eighties, but ultimately decided against it when he saw some of the complications that arose. "I could have specialized in *repair* of laparoscopic injury," he told me. "There were that many." Kagan (who happens to be my father-in-law) tells me that, among other problems, pushing an instrument through the abdominal wall instead of making an incision requires a fair amount of force, which creates the risk of spearing some innocent organ or blood vessel.

Recent studies seem to bear out Kagan's concerns. Researchers writing in the April 29, 2004, issue of *The New England Journal of Medicine* found that while men who had had hernias repaired using laparoscopic surgery had less pain after surgery and got to go home sooner than those with conventionally repaired hernias, they also had twice the number of recurring hernias and more complications. Two patients in the study group actually died after laparoscopic surgery (no one in the study died after conventional surgery). The study's conclusion: if you need to get a hernia repaired, it is better to go with conventional surgery, even though it'll be more painful.

The demand for laparoscopic surgery continues to grow, even as new evidence of increased complications from this type of surgery comes to light. A study published in the May 10, 2008, issue of *The Journal of Clinical Oncology* compared laparoscopic versus conventional, or open, radical prostatectomies—a treatment for prostate cancer that is increasingly performed laparoscopically—and came up with results similar to those of the hernia study. Patients went home sooner after laparoscopic surgery, but were at greater risk for complications from surgery. One in four patients required follow-up radiation or chemotherapy, versus only one in ten of open surgical patients whose surgery, in effect, didn't work. Contributing to the decreased effectiveness of laparoscopic surgery, one surgeon commented in *The New York Times* after reading the study, may be the fact that a surgeon operating laparoscopically can't feel the cancerous growths, and thus has a harder time getting them all out.

Leaving aside the possible risks to patients created by punching holes into them instead of slicing them open in the manner of conventional surgery, the trouble with laparoscopic surgery from the *surgeon's* point of view was that it required him or her to operate in an extremely awkward manner. He had to push his instruments down into the bowels of a patient, where he couldn't directly see anything, and watch a TV monitor to see what he was doing. But the position of his hands and the effects of his motions didn't precisely correlate with what he saw the instruments doing on the screen. In fact, what the surgeon saw on the screen was the mirror image of what he was actually doing—if he moved his instrument to the right, the image of that instrument on the screen moved left. This was disconcerting, to say the least, and surgeons had to undergo months of specialized training to regain something of their natural dexterity in such unnatural conditions.

On top of that, there was the mechanical difficulty imposed by the need to work the manipulators inside the patient's body without putting undue pressure on the sides of the access ports. A robot would be ideally suited for keeping absolutely steady the tubes through which it extended its manipulators into the patient's body while slicing and sewing inside, and it would be simplicity itself to program a robot to present a corrected, non-mirror, image of the operating field to the surgeon. All of which presented an opportunity for a spunky new medical robot looking to get its start in the world.

By the mid-1990s, with the number of cases burgeoning, the laparoscopic surgery community was in desperate need of some help, and SRI's Medical Systems and Devices division finally had the business case it had been looking for. In 1995, SRI's Moll helped establish a new company called Intuitive Surgical, to build on SRI and DARPA's work and create a commercial system—one uniquely specialized for laparoscopic surgery. Instead of working on a patient remotely as in the SRI system, the surgeon would sit just across the room, manipulating instruments in the patient's body that were small enough to fit through the minimally invasive ports and using the camera view to watch what he was doing in the way that was most natural—looking down at his hands just as he would when performing conventional surgery.

The result was the da Vinci Surgical System, with seven degrees of freedom. SRI fostered development of this new surgical robot, built the early prototypes, helped Intuitive Surgical get venture capital, and in 1999, the year Satava left DARPA, the company moved out to its own headquarters in Sunnyvale, California, and introduced its new product to the market. Today, the $1 million da Vinci robot is the gold standard for robot-assisted minimally invasive surgery. No commercial competitor even exists, and it's rapidly gaining users not only throughout North America but all over Europe and in India, Australia, Singapore, and Saudi Arabia as well. In 2008, the da Vinci's customer base of 850 hospitals around the world was growing at an astonishing rate of 25 percent a year.

Good for Intuitive, and ultimately good for SRI's bottom line. But Low couldn't let go of the dream of surgery by true telepresence, with a system that would allow a surgeon to operate on a patient from anywhere in the world—or beyond. Fortunately for Low, Satava returned to DARPA in 2002 and picked up the Trauma Pod project where he had left off.

In its latest iteration, the phase one setup that I saw at SRI, the

Trauma Pod sought to demonstrate all of the technologies needed to care for a critically wounded soldier in the battlefield without a surgical team at his side. Phase one, launched by Satava in 2005, didn't concern itself with cramming all of that new technology into the kind of rugged easily transported container that would eventually, if all went well, see service on the battlefield. That would wait until phase two or three. Phase one would be difficult enough. For that, Satava wanted not just a telerobotic surgeon, but an entire robot surgical team.

"It's important that we don't have a surgeon there," Low told me of the Trauma Pod concept, "but if you've got a robot and a whole team of nurses and an anesthesiologist out on the front line, it's not really very practical." Satava was determined to take the technology to the next level, Low explained to me during my visit. "The question that Rick asked about in phase one was 'Can the whole thing be done with robotics?'" That meant, Low told me, building "not just the robotic surgeon, but a robotic scrub nurse, a robotic circulating nurse, and a robotic anesthesiologist, and a robotic—you know, you name it." The goal for phase one of the Trauma Pod was just to try to accomplish the basic task first. Reducing it all to the size and weight parameters of a field-ready system would wait. "If you can't do that without limits on size, weight, power," Low recalled of his direction from Satava, "then we're just wasting our time." Phase one of the Trauma Pod would culminate in a demonstration of some procedure that might have to be performed on a battlefield to save a soldier's life if he couldn't be brought to a field hospital immediately, with no human in the operating room. No one, that is, besides the patient.

Led by Pablo Garcia, SRI's principal investigator and engineer for the project, the team hit all their marks to demonstrate their system within the two-year time limit imposed by Satava. Satava himself had left DARPA again in 2006 to become the senior science advisor at the U.S. Army's Medical Research and Materiel Command (MRMC) at Fort Detrick, in Maryland. Nevertheless, in March 2007, the SRI team and DARPA officials, including Satava's successor as program manager, army Colonel Geoffrey Ling, along with, Low told me later, "about fifty other people, all kinds of brass," gathered in the room in which I saw the Trauma Pod surgical suite and its rubber patient. "It was like a synchronized dance. It was quite a show," Low recalled with some satisfaction.

When I saw it, the operating room—or, rather, operating theater—was just as the team had left it for that demo a year and a half before my visit. SRI was waiting for DARPA to continue on to phase two, and all the robots stood poised and ready to go into action at a moment's notice. The last that Low and his team had heard, DARPA had signed off on the funding. Now everyone was just waiting on the army.

When DARPA director Tether came into office, he wanted to make sure that every project under his watch had a clear path to transition from DARPA's R-and-D program to completed device, system, or set of technologies that would actually be used by one of the armed services or in a commercial setting that the services could easily avail themselves of. That's why he told Ling, the Trauma Pod's new program manager, that he couldn't proceed to phase two until he had a customer in the services. Ling found that customer in the form of the MRMC. It was up to the army, then, a far more bureaucratic organization than either DARPA or SRI, to put together a plan to take over the Trauma Pod after phase two was completed.

"It's a big, important program," Low told me, "and we'd like to move forward."

The room containing the robotic surgical suite and its extremely patient patient was divided in two, with the operating room on the left as Low and I walked in. The operating room was demarcated by glass partitions etched with the stylized world map of the SRI logo on them. A da Vinci surgeon's console squatted in one corner, just outside the glass doors. Across the room, in the opposite corner, also just outside the OR, was a more traditional computer workstation, with flat-screen displays. To our right as Low and I entered the room was the group's latest effort, the M7 surgical robot—like its M4 predecessor, named for its number of degrees of freedom—along with its surgeon's console.

The patient lay on a life support gurney built by one of SRI's partners in the project, Integrated Medical Systems. The gurney, a commercial version of Satava's LSTAT, had been to provide oxygen, IV fluids, vital signs monitoring, and other emergency functions to a soldier being medevacked from a battlefield. Here it performed one more vital task: it provided a stable platform for an onboard optical scanner mounted on a sliding black arm bent over the patient. Once the LSTAT was situated in the operating room, the scanner would slide down the length of the table, capturing a three-dimensional image of the patient so that the Trauma Pod robots would know where his body began and the rest of the equipment ended.

All of the robots in the room had the ability to orient themselves to each object they expected to find there. They could be set up in any operating room, and once activated, they reached out their grasping arms to find specially shaped lugs on the other equipment. Low moved into the room ahead of me and around the machines, to show how the robots would create internal maps of their positions relative to one another. "So the robot comes over and grabs here, and feels this," said Low, grasping a squarish lug on the face of the supply dispenser to the patient's left, just as the robot scrub nurse between it and the gurney would. "Okay, now I know where this is. It grabs here and here," he continues, touching other lugs in the equipment. "And it comes over and touches here and here, and now it knows where this is." Like blind people, the robots would orient themselves by feel, and then move confidently as long as nothing budged them from their places. "Each of these objects has a lug, and as long as you're within a few centimeters, [the robot will] close on and it'll store where that is." In other words, technicians could set up the robotic operating room wherever it was needed just by placing the robots in the neighborhood of where they expected to find one another.

The central computer coordinating the movements of all of the robots also tracked their precise positions and trajectories. "If, for example," Low explained, "the surgeon moves his hand in such a way as to cause the da Vinci to lurch out into the path of the scrub nurse, that's immediately detected, and the path of the scrub nurse is automatically recalculated in real time so that they don't crash into each other." He paused a moment. "I haven't actually tried it, but I think if you go like this to it"— here he lunged at the scrub nurse as he might do while using the da Vinci at the heart of the Trauma Pod—"it kind of dodges out of the way." I had to laugh at that, imagining a bunch of engineers whooping it up after hours at an office party and playing with the machines.

The scrub nurse, built around a Mitsubishi PA-10 industrial robot by Oak Ridge National Laboratory and running software by the University of Texas, was known around the SRI lab as the SNS. SNS originally stood for Surgical Nurse System, but the robot is now referred to only by its initials, in deference to human nurses who might fear for their jobs. A single articulated arm with two pincer-like "hands" on the end of it, the SNS looked like a scaled-down factory robot of the kind used to assemble cars.

The actuators and motors controlling the da Vinci system, and from which four robot arms hung poised over the patient, stood to the patient's right. As with all da Vinci systems, the surgeon's console outside the glass partition contained the da Vinci system's only computer processors, and it had to be in the vicinity of its slave unit for the system as a whole to function. No chance of remote surgery with this setup—the two sections of the system were hardwired together. Not that the da Vinci was even qualified to do the kind of trauma surgery the Trauma Pod engineers and managers had in mind. "It's for doing, you know, heart bypasses," said Low, "prostate removals, things like that. But it's the only commercially available robust robot. It's a placeholder, really."

Even so, the team had been able to give the da Vinci enhanced capabilities via a software upgrade aimed at letting a single surgeon control all of the robots in the suite. "He's not just responsible for tying knots anymore," Low explained to me of the human surgeon. "He's orchestrating the entire OR." To figure out how best to enable the surgeon to pull off that daunting task, Low and his team sought the advice of working surgeons. "We studied how surgeons interact with nurses and technicians," explained Low. And they had surgeons test-drive the robot surgery and report what worked well and what didn't. The result was a system with which a surgeon could interact just as he would if he were actually present with live nurses—through verbal commands (speech recognition was one of the new capabilities the team had given their da Vinci system) and gestures. He would request the instrument he needed, and the robot scrub nurse would fetch the appropriate instrument from a rack and tell him verbally that it was ready. At that point, the surgeon lifted his "hand" from the patient, just as he would if he were actually present in the operating room. That signaled the robot scrub nurse that he was ready to accept the instrument. Once it got the signal that the surgeon was ready, the SNS took hold of the appropriate arm of the surgeon's robot, detached the instrument the surgeon had been using, placed it in a receiving tray for possible later use, attached the new instrument, and told him he was all set to continue working.

"Debakey forceps, right hand," the surgeon might say.

The SNS whirs into action, moving to its right to pick up the forceps—essentially a pair of tweezers at the end of a long, thin rod—from a stand to its right called the tool changer. With the tools on board equipped with the same kind of radio frequency ID chips used to inventory merchandise in retail stores, the tool changer can keep track of all of its gear; it spins to present the requested tool to the SNS. Tool in hand, the SNS swings into position beside the surgeon's robot. "I'm ready, sir," it tells the surgeon.

The surgeon lifts the arms of his robot from the patient, signaling that he's ready to take the new instrument.

The surgical system then takes control of the surgeon's arm to move it, under computer control, away from the patient and present it to the SNS. The SNS grasps the old tool with its free hand, detaches it from the surgical robot, inserts the new tool with its other hand, and tells the surgeon, "The tool is ready," as the computer passes control back to the surgeon.

When the surgeon requests expendable supplies, such as sutures or gauze, the SNS retrieves the appropriate supply from yet another system.

Behind the SNS, a big glass-walled cabinet is filled with specially packed expendable supplies, preselected by the project's surgeon consultants based on the procedures the system would likely be called upon to perform. The supply dispenser, built by General Dynamics, and of which the cabinet is the main part, constitutes a robot all of its own. When the surgeon orders a particular item, the dispenser's own arm inside locates the correct packet of sterile supplies from a series of racks and delivers it to a slot on the outside of the dispenser. From there, the SNS takes it and delivers it to the surgeon's outstretched robot hand. After the surgeon has finished using that particular set of supplies, the SNS moves it to a rotating tray called a fast cache, on its left, where it can quickly retrieve it again if the surgeon requests it.

As an added bonus, the system automatically keeps close tabs on all the tools and supplies, ensuring that everything is accounted for at all times and nothing gets left inside the patient. "How many sutures do I have left?" the surgeon might query, to which the system would verbally give him a precise tally. Even more important, "How many sponges are in this patient?" the surgeon might ask, if he's lost track, to which the system would give him a swift and accurate reply.

A precise accounting of supplies and instruments left in the patient isn't so critical in the Trauma Pod's first intended operating environment, a battlefield, where the mission will simply be to stabilize a patient just long enough for him to be opened up again within an hour, and so a stray sponge or two won't matter. But in a more traditional hospital environment, keeping track of sponges or instruments that might get accidentally left inside a patient is a serious business, and one that all too often gets flubbed, to the later dismay of patients—and to the surgical staff that patients have been known to take to court for damages.

The phase one Trauma Pod system operates as a seamless whole under the direction of the surgeon. Getting all those components, each built and programmed by a separate Trauma Pod project partner under SRI's direction, was the major challenge for Low's engineers. The engineers had to define every aspect of the surgical suite's operating environment: not just the physical layout of the surgical suite and the positions of all the robots and how much space they would have to work within and the fact that there would be a central computer coordinating the actions of all of the robots, but also the computer language in which all of the components would communicate with the computer. All of the commands that each piece would receive had to be defined, along with all of the appropriate mechanical and electronic responses. Essentially the team had to invent a new computer operating system just for this application. "You can't expect someone to deliver this in a month," said Tom. Indeed, just that process of defining the environment the robots would operate within took a full four months of the two-year project.

Once that environment had been defined, each of the contractors sent a computer to the team at SRI loaded with the software that would drive the finished robots. The team installed the computers in a rack and plugged them into the supervisory system. The software in the component computers could then drive virtual robots in a virtual environment before the hardware was even finished, and give consulting surgeons a chance to help define how their colleagues would use the finished system—which is how the SRI team came up with the process by which a human operator would control the system with gestures and voice commands.

For phase two of the Trauma Pod project, the team planned to narrow the robot surgical suite's focus to four distinct procedures that could save a critically wounded soldier's life. Low ticked them off for me: "Airway management. Penetrating chest injuries. Some pneumothorax [i.e., collapsed lung] and incompressible exsanguinating hemorrhage [that is, a pumping wound that can't be squeezed closed]. Simple. And head trauma's another one that's important." Though that type of injury wouldn't be treated by the Trauma Pod. Instead, the system's job in that case would simply be to identify it as such.

With the challenges defined, the engineers could break down the quick patch-ups for these types of injuries into a series of mechanical actions that could be performed by a robot. But perhaps the greatest challenge of all had yet to be tackled. "One more thing that the army's asked for," Low told me, "is autonomy." In other words, the system would have to perform on its own, without the kind of direct human input required by the da Vinci system, and hence the phase one iteration of the Trauma Pod. "On the battlefield," Low explained, "you don't necessarily have the ability to send high-definition video back and forth to surgeons. Or have low-latency, high-bandwidth communications. So it's imperative that the system ultimately be able to operate—at least do a lot—without that communications channel." Without, in other words, a human in the loop for more than a quick check now and then as conditions allowed, or to perform the grossest of tasks, such as loading the patient into the Trauma Pod. "Our view is that if someone who's not medically trained—a buddy or a medic perhaps, but most likely another soldier on the battlefield—anything they can do without training, they should do. So, putting the patient on the Trauma Pod stretcher: you don't need a robot to do that. Taking off the body armor—huge problem for a robot, something you and I could do without being taught."

Low and his team wanted to use their robots for what they did best, Low told me, "and that is going to very precise coordinates repeatedly and doing it with guidance from some external system such as a CT scanner." With all the systems in the Trauma Pod working in concert under the control of the supervisory system, the robots could find a shredded artery and patch it long enough for the patient to survive the trip to the hospital. Low even ventured that in some cases his robots might be able to perform a given task not just as well as but even better than a human surgeon. "All the coordinate systems of all these different pieces are locked together," he explained, "so that the CT scan says it's XYZ where we have to go. The robot can go there, and even if there was a surgeon there with that information, he couldn't do it."

In contrast to phase one of the project, Low told me, "The vision for phase two is not to duplicate what a surgeon does with his hands, but to most efficiently address those four needs." In other words, the goal was to create not an all-purpose autonomous surgical suite—although that might be possible someday in the more distant future—but a working system on the kind of accelerated schedule required by DARPA director Tether. "We're not trying to do everything," said Low. "We're trying to do four things really, really well in a package that has promise to be made small enough and light enough to actually go out into the field."

Step one in any procedure for treating any of the four types of injuries Low's team had identified would be to put in an access line. To do that, the system would need to find an appropriate blood vessel and put in a needle or catheter, without human direction or control. Low explained the procedure to me rapid-fire, taking the point of view of the machine. "So, ultrasound imaging, identify where a major vessel is—maybe it's the iliac, the aorta, a femoral artery, something like that—and get a needle into that, and then an expander and a catheter. And now I've got access. I push in contrast-enhancing radiographic material and then I use a new generation of CT scanner that will be developed in phase two of Trauma Pod that can image the soft-tissue body and identify the major injuries."

That's yet another device that will have to be designed, built, and tested for the project, and just that piece alone could bring significant benefits to the field of trauma medicine, and not just on the battlefield. "The difference between that and the clinical systems that are used today in hospitals is that I don't need to image a one-millimeter cancer tumor anymore," Low explains. "I don't care about that. I'm looking for the chunks of shrapnel and the torn aorta." Dialing back the imaging power of a conventional scanner necessarily means the device will consume less power. Engineers would also be able to shrink the room-filling scanners used in hospitals to machines the size of a suitcase, as well as greatly increase their operating speed. The result would be an imaging tool that was formerly the exclusive domain of well-equipped hospitals, that could now be carried on an ambulance or medevac helicopter as well as the Trauma Pod, and that could help make spot diagnoses and more effective emergency treatments at the scene of car accidents and other places where patients might be injured.

In the case of head trauma, the Trauma Pod system wouldn't be able to do more than transmit the diagnosis to the nearest field hospital to find out if the hospital had the staff to take on a head trauma case, or if the victim should be rerouted to another location better equipped to handle him. But for a collapsed lung case, the Trauma Pod could treat the victim en route. "Let's say I do the scan," says Low, "and I see that the guy has a collapsed lung because he had a big piece of shrapnel go through his chest. I have to reinflate that lung. I'll take a needle. I'll find the ribs with ultrasound and CT image, I'll pop the needle in, pull the air out, stick a patch over, fixed."

The standard method of managing a patient's airway—that is, making sure there is an unobstructed path between the lungs and outside air, including keeping blood or other fluids out of it, and allowing for mechanical ventilation in cases where the patient can't breathe on his own—is a process known as intubation. To intubate a patient, doctors or nurses push a tube down the patient's throat, through the larynx, past the vocal cords, and into the trachea, where it is secured in place by inflating a balloon-like device near the tip.

"Really hard," asserts Low. "Even in the OR and the ICU, they have trouble doing it with trauma." Fortunately, there's a simpler method for trauma victims, one that can be used by medics in the field, though still not without difficulty. The so-called laryngeal mask airway consists of an inflatable seal and a tube that goes down just past the back of the mouth to the pharynx. "We want our robot to be able to put that kind of mask airway into place," says Low. "Same kind of thing—imaging, using the robot to put that in place, and lock it so we get airway. In the event we can't get that mask airway in place, we'll do a cricothyroidotomy," i.e., punch a hole in the patient's windpipe and insert the breathing tube there. As with gaining access to a blood vessel, it's a matter of the robot guiding a needle to a precise set of coordinates.

The final procedure on Low's list is hemorrhage. "And," he says, "this is the really wild one." Wild in the way the robots would tackle the problem—not at all the way a human surgeon would, but just as effectively, perhaps even more so, according to Low. With body armor protecting the torsos of American soldiers in Iraq and Afghanistan, the IEDs that have claimed the bulk of the casualties do so by blowing off limbs. And when, for instance, a leg is torn off too high for a medic to get a tourniquet around what remains of the limb, a wounded soldier can quickly bleed to death. What's called for, if the victim can be brought to a hospital in time, is for a surgeon to reach inside the wound, find the artery that's pumping the victim's blood out of his body, and clamp it off. The Trauma Pod will use a radically different approach to do the job.

Using the suitcase CT scanner, the robots will find the bleeder before going in. Once it is located, it will be a relatively straightforward matter to push a catheter in through the blood vessel access the robots will make as soon as the patient enters the pod. To patch a leg wound from there, the machine will thread the catheter through the aorta, and then down the iliac, or other bleeding blood vessel in the leg, to inflate a balloon at the catheter's tip from inside the vessel to seal off the leak. Find the coordinates, put together a plan, reach the coordinates. All actions ideally suited for robots.

With all of these procedures, says Low, "we're just trying to buy thirty minutes, an hour on the outside." Again, just enough time to get a critically wounded soldier to a field hospital. "I wish we could say we could save them all," he told me. "We're not going to be able to." But for those whose injuries can be effectively patched on the go by robots acting autonomously, the Trauma Pod might just mean the difference between a swift death by bleeding or suffocation and enough time to get life-saving treatment by human surgeons.

The trip to the hospital will necessarily be a bumpy one, ideally in a helicopter rather than a truck. "Ultimately we'd like to be small enough to fit in a Black Hawk [helicopter]," said Low. And the system won't do much good if it has to sit on the battlefield, an easy target, while the patient is diagnosed, patched up, and stabilized *before* taking off. No, to be most effective, the thing will have to get to work on the patient while the Trauma Pod is in motion, first lifting off, and then flying to a field hospital, possibly even while the chopper's pilot takes evasive action to avoid enemy fire. Which will present yet another challenge: keeping the robots' hands steady in the patient's body even as the machine itself shakes, rattles, and rolls.

To solve that problem, the SRI team had already developed, by the time of my visit, what they called an acceleration compensation system. The system relied on accelerometers—sensors that detect motion—to analyze in real time the motions of the system at any given moment, and command a robot's hands to shake and jitter in precisely the opposite directions in all three axes to hold them steady relative to the patient.

The system got a workout in December 2007, on board a C-9 jet in a flight sponsored by NASA as part of SRI and the University of Cincinnati's work on telerobotic surgery for future space missions. Researchers took turns sitting in the back of the airplane and working the controls of the team's latest surgical system, the M7 robot, as the airplane climbed and dove to induce alternating periods of two-g and zero-g flight. The operator, his hands strapped down to stabilize them, cut and sutured squares of flesh-colored latex while the M7 duplicated his motions precisely, accurately correcting for the motions of the airplane.

To let me get a feel for how the M7 robot performed, Low sat me down at the M7 console there in the room with the surgical suite and instructed me to put my eyes to the binocular-style viewer and place my hands on the controls on the table just below.

In contrast to the da Vinci system at the heart of the Trauma Pod phase one system, with its hardwired connection between human operator and machine slave, control for the M7 passed from the master console to the slave robot hands via Ethernet, the standard computer networking protocol used in homes and business everywhere. Bringing Low and company's vision for telerobotics full circle, this enabled the system to be controlled remotely from anyplace where both slave and master had access to an Internet connection, wired or otherwise.

Looking through the binoculars, I saw the operating arms below me in 3-D, as though I were looking at my own hands. The illusion was perfect. It really seemed to me as though I were looking down at instruments that were right in front of me. In fact, the arms were behind me and to my left. The stereo image I saw was fed to the dual screens in front of my eyes from a pair of consumer-model camcorders peering down at an ersatz operating table with the precise separation needed for three-dimensional vision. The table was well lit, with bright though not glaring lights. I could see perfectly as I moved the robot's arms through the air above a square of latex resembling a patch of flesh on a patient—the same type of tissue model used on the zero-g flight a few months before. A scalpel and a hooked needle and sutures (i.e., surgical thread) invited my robot-assisted manipulation.

When I grasped the suture in both of my pincer "hands" and tugged experimentally, I felt resistance, part of the haptic awareness built into the system. The sensation was such a realistic reproduction of my everyday experience, and hence only on the threshold of my consciousness, that I wasn't even certain I was actually feeling it until I closed my eyes to shut out the visual input.

Then I could tell that the haptic sense actually wasn't all that strong, giving me feedback in a dulled, distant kind of way. The effect was as though my hands were somewhat numbed. I could feel some resistance as I dragged the manipulator over the surface, and then a sudden loss of that resistance as the manipulator dropped into the incision. I could barely feel any resistance at all as I cut into the flesh with the scalpel (which Low helpfully placed in my right manipulator). I felt the resistance most strongly when I pushed the square of flesh with each manipulator so that I was pushing the two manipulators together with the square of flesh in between—pressing the flesh, as it were.

I found that I could not reliably "operate" by feel alone. The system's haptic sense, while certainly helpful, did not give me the *tactile* feedback that I would have gotten through my actual hands. After I prodded the flesh with my left manipulator, Low suggested I then try poking the table. The somewhat spongy resistance I felt with the flesh became a more solid-feeling "thump" on the table, a sense of an implacable obstacle. Even without the tactile sense, the illusion of using my own hands to manipulate something across the room or half a world away was uncanny.

A clutch under my right foot allowed me to stop the system from reading my movements. When I disengaged the clutch by lifting my foot, I could reposition the controls to get my hands in a more comfortable position, or to move back from a hard stop at the limits of the controls, without moving the manipulators poised over my patient at the same time. When I was ready to work again, I simply pushed the clutch in to reengage the system. The effect was like picking a computer mouse up from a desk to put it in a better position to work with it, as I might do when I ran out of table or bumped against a desk lamp or other obstacle. Something I could not do with a mouse that I could do with the M7's controls was to move in three dimensions, as I did when I rotated my wrist and the machine responded accordingly. The wrist rotation was but one of the seven degrees of freedom of which the M7 was capable.

The SRI team was already at work on its next challenge for telerobotic surgery—robot surgical systems for use on long-duration space missions. The group had received funding from NASA with a view toward developing a system that could be used by astronauts on a space station and even on a future voyage to Mars. The system developed for a space station would be a relatively straightforward application of SRI's telerobotic surgery and motion-compensation technologies to cope with the microgravity environment of space. Given the close quarters of a space station, the astronauts would have to break the system out of storage when it was needed, and assemble it from pieces.

Low was realistic about the kinds of procedures the system might be called on to perform. "To be candid," he told me "if the guy gets crushed between two structures as he's putting together the space station it's not a very good scenario, period." But the autonomous surgical system could become very useful in treating certain illnesses caused by extended-duration weightlessness. People living in microgravity lose muscle and bone mass in a process similar to that experienced by patients on extended bed rest or who are unable to get up from bed because of paralysis or other debilitating conditions. Astronauts can keep from losing some muscle and bone by using specially designed exercise bikes and other equipment that depends on resistance rather than gravity to give them a workout. But after weeks and months in microgravity, some muscle and bone loss is inevitable. Also, calcium from the bones can build up in the bloodstream and create kidney stones and other problems. A kidney stone that becomes lodged in the ureter, the duct that conveys urine from the kidneys to the bladder, can create a pus-filled abscess that a robot surgical system would be ideally suited for finding, piercing with a needle, and draining to relieve pressure. Until the astronaut returns to Earth and has the abscess properly treated, the pus will continue to build up, requiring repeated lancing and draining, perhaps every two weeks, but even that minimal treatment could be all an astronaut needs to continue working to fulfill the mission.

In the case of a Mars mission, a communications lag time caused by the extreme distances involved will grow longer and longer the farther a spaceship travels from Earth. Mars averages four light-minutes from Earth, meaning that those wonderful stereo views of a patient transmitted by some future spaceborne version of the Trauma Pod would take that long to reach an Earth-bound surgeon, and then that long again for his or her motions, translated into radio signals, to reach the pod. Eight minutes for two-way communication is certainly far too long for a master-slave system to be useful for surgery, so the system would have to function perhaps even more autonomously than on a battlefield.

Low and his team envision mission planners ordering head-to-toe CT scans of each astronaut, and loading those into the Trauma Pod's computers along with preprogrammed surgical procedures. The procedures could be modified to fit specific injuries that might befall the astronauts in flight, and the data from the scans made on the ground could be updated with diagnostic scans on board the ship, to give the robots the best possible chance to minister to their human charges.

In addition to running experiments on zero-gravity parabolic airplane flights, the SRI team has also installed the M7 (along with another experimental surgical robot from the University of Washington called Raven) in an undersea habitat called *Aquarius*, operated by the National Oceanic and Atmospheric Administration and used by NASA as part of the Extreme Environment Operations project in the Caribbean, off the Florida Keys. In 2006 and 2007, University of Cincinnati surgeon Timothy Broderick, along with his fellow "aquanauts," NASA astronauts training for space missions, descended sixty feet to the ocean floor with the M7 and Raven for up to two and a half weeks at a time, to practice techniques for living off the planet. Receiving instructions over a wireless Internet link, the M7 performed beautifully in a series of suturing and cutting tests, Broderick told me.

Of course, other, non-space-related challenges asserted themselves during this series of tests—such as getting the surgical robots through sixty feet of water to *Aquarius*'s so-called wet porch and into the dry interior without ruining their delicate electronics; or getting their hard disk drives to spin in the two-and-a-half-times-normal atmospheric pressure maintained in the habitat to allow the aquanauts to venture outside for long periods without developing decompression sickness. But all of this was relevant research, for space missions were by no means the only possible context for robot surgeons. And if Satava and the other visionaries have their way, robot surgeons operating in inhospitable environments such as war zones, outer space, and underwater will be just the beginning. For starters, Low and the team at SRI would like to see robots rolling around in the homes of the elderly and others who might need extra help each day, sorting pills and making sure the oven is off, for example. The robots would be controlled remotely, giving caregivers remote eyes, ears, and hands inside the homes of the people they were looking after, no matter how far away they were.

For his part, Satava still sees the greatest potential for his robots in the operating room. "Everybody's using a word processor and we are using a typewriter," he told me after my visit to SRI. "By that I mean, when you do a document, you edit the document until it's perfect, and press the button, and it prints. So we're going to use that analogy. If we take the body scan of the individual, operate on that, make mistakes, correct the mistakes, and when the surgeon is happy with the perfect one, then he will send that to the robot and the robot will do it for him without error."

In Satava's vision of the future, which he continues to work to bring about in his position as the MRMC's senior science advisor (read: "resident futurist"), surgical patients would assume a role little different from automobiles on today's assembly lines. As part of the diagnostic and intake process, a doctor would give his or her patient a total body scan to upload a computer model of the patient into a computer system. This Holographic Medical Electronic Representation, or HOLOMER, "is the same as a CADCAM of, say, a car or a pump or something like that," Satava told me. The surgeon would operate on the HOLOMER of the patient rather than on the patient himself, correcting mistakes, editing and perfecting, reviewing simulations of the operation, and then only when the procedure was perfect, the surgeon would send for the actual patient, load him into the automated operating room, and click Print— that is, send the command to the robots to get to work.

The advantage to using robots regularly for surgery like that, says Satava, is that robots can operate ten to fifteen times more accurately and eight to twelve times faster than a human surgeon. "Can you imagine a trauma operation taking five minutes instead of an hour?" Fulfillment of that dream, he estimates, is some fifty years off. A lot of the big challenges are still ahead—greatly increasing the autonomy of the robots to the point where they are capable of performing far more than the four basic procedures envisioned by Low and the SRI team, for one, enhancing the resolution of the total body scan to create the HOLOMER, for another. "There is some technology that we would have to invent," Satava admits. But he believes that most of the needed technology has already been developed. "It's just a matter of putting it together and engineering it and proving it works."

Andrew Kagan, the surgeon who believes laparoscopic surgery is overused, has his doubts. "You can't treat all living tissue the same way as you would inert material," he insists. Human beings breathe and pump blood, and their organs don't hold still for automated procedures that would rely on body scans.

Will robots ever be able to adequately compensate for the myriad ways our bodies fail to conform to static computer models? Perhaps we'll find out in the next fifty years. In the nearer term, the DARPA-funded medical robotics technologies now in development may just give trauma patients who have no other options a fighting chance.

BACKSEAT DRIVERS

I DISCOVERED that although DARPA's doors were firmly shut to outsiders, this wasn't always the case with its contractors. I was able to establish direct relationships with the researchers at SRI, Johns Hopkins University's Applied Physics Lab, and other places doing advanced research for DARPA, without actually getting the agency's help to do so. That was all to the good, but to get access to DARPA's program managers—the brains behind the operation—and to the director himself, the hands-on manager who directly guided the work of the program managers, or to the agency's unremarkable-looking office building in Arlington, Virginia, I had to get past one woman.

Jan Walker, DARPA's public relations manager, holds the keys to DARPA in a tight fist. She handles the e-mail queries, the phone calls, and all the attempts by journalists to speak directly to the program managers and the agency's director and to find out just what the heck is going on in that office building in Arlington. Walker is not subject to the term limits imposed on the program managers and the director, and she has been there longer than any of them. She came to the job from a public affairs office at the Pentagon some dozen years before I started my research in the summer of 2007. In many ways, she was the perfect woman for the job—pragmatic, and unromantic about the job of communicating to the rest of the world the exploits of America's secret weapon. "Did you have any special interest in public relations before this?" I once asked her. "No," was her unadorned reply.

Like any successful organization with any sort of public presence, DARPA recognizes the need to fulfill a public relations function. It has to tout its accomplishments a certain amount to win the funding it requires from Congress and also the brain power it needs to do its work. But as befitting a military agency that keeps half of its focus on secret programs, it has a natural aversion to too much attention. Hence the need for a gatekeeper such as Walker, who is happy to keep the flow of information leaving the building at a trickle.

Naturally one of my first moves was to phone a couple of program managers directly; their direct contact information is listed on the DARPA Web site after all. True to his maverick sensibilities, one of them actually did speak with me at length, after first admonishing me that he wasn't allowed to, and so of course couldn't be quoted. He gave me some excellent suggestions for programs to cover in the book and the names of some of the researchers I could speak to without DARPA authorization. But there was no escaping the fact that to tell the story of DARPA, I would have to go through Walker for some real access to the people who continued to make it one of the prime engines of innovation in the world.

Walker was initially noncommittal about the possibility of the agency supporting my book project. I got my chance to meet her face-to-face at the Urban Challenge, a DARPA-sponsored robot car race held in the Mojave Desert during ten days in October and November of 2007. The press was invited, and Walker would preside over the media tent. With an assignment in hand to blog the contest—all ten days of it—I flew out from New York as much resolved to report with gusto on the world's most unusual auto race as to make myself an indispensably cheery part of each of Walker's days in this desolate corner of California and win my way in the door at DARPA.

.   .   .

OCTOBER 25, 2007, Victorville, California. The big former air force base now known as the Southern California Logistics Airport sits smack in the middle of nowhere. It's the perfect place to hold a robot car street rally, far from the potential for collateral damage.

As I drove in for the kickoff for DARPA's Urban Challenge, the control tower beamed a searchlight like that of a lighthouse through the early morning haze. Well-maintained runways and airplane hangars quickly gave way to streets of empty barracks and disintegrating suburban-style homes. The abandoned neighborhoods were dotted with dead trees. There must have been grass here once, along with the trees. Maybe the buildings were once painted different colors, too. But the houses, the trees, the now-sandy ground, had turned the same dun color as the surrounding desert. All gave the impression of purposeful neglect rather than outright abandonment. The ghost landscape didn't seem so much dead as animated with a different kind of life than when these neighborhoods sheltered humans.

I followed a sign pointing the way to the Urban Challenge and passed a staff parking lot enclosed with new chain-link fencing and bearing prominent signs warning everyone but staffers away. After that came the DARPA command post, which consisted of three connected mobile offices surrounded by more fencing and patrolled by unarmed guards. Inside, during the race, controllers—for the most part, DARPA program managers borrowed from their usual duties—would be able to monitor each of the cars, tracking its progress on an overhead map. They would stay in contact with other program managers out on the course, secured at observation posts behind concrete K-rail barriers, and of course with the chase drivers, each of whom would follow one robot car. Video feeds from points along the course augmented video that would be sent from an airplane circling overhead. On the roof of one of the bungalows squatted a military green mobile satellite receiver, sandbagged in place. On the asphalt to one side, a portable communications tower bristled with satellite dishes.

The kickoff meeting took place in a circus-size white tent with a big

Welcome sign on the outside. There, Tony Tether greeted members of the thirty-five teams gathered to compete for a $3.5 million purse in the world's first autonomous vehicle race through city streets. Admittedly these were to be carefully controlled city streets, devoid of any innocent pedestrians, ordinary human traffic, or even functional buildings that might need repair after a car went on an uncontrolled rampage and plowed into them. Still, this was the first time that robot cars would attempt to navigate streets while keeping to the correct side of the road, stopping at stop signs, signaling their turns, yielding to oncoming traffic, and otherwise adhering to the driving laws of the State of California—all without any human input. The idea was to demonstrate the ability to carry supplies and equipment through urban war zones without putting humans at risk. The cars would simply be pointed in the right direction and told by their humans to *go*. The first car to complete the sixty miles of the race course within the allotted six hours would win for its team the first prize of $2 million, while the second and third place winners would get $1 million and $500,000, respectively.

Tether took to the stage in the circus tent and, with the help of some presentation slides, outlined the rules of engagement for the Urban Challenge qualifying event. There was to be no alcohol in the "pit" area. "Don't go into the bombed-out buildings," he said, referring to the abandoned and crumbling houses and barracks on the base. "There are security people everywhere, twenty-four/seven, and they're just waiting for you to do that. So don't do it. You don't want the government to have to get involved. Because when the government gets involved, it gets nasty." There was to be no autonomous driving on real streets or other areas with live traffic. And the most important rule of all, said Tether, "Don't hit anyone!!!" This last directive was presented on the slides in letters three feet high, with three exclamation points, just to make sure no one missed it.

The actual Urban Challenge race would be held a week and a half hence, on November 3. Between now and then, DARPA judges would use the National Qualifying Event, or NQE, to weed out any cars that displayed dangerous tendencies. They planned to reduce the number of competitors from thirty-five to a more manageable twenty.

Step one: make sure that each car could be reliably stopped in a hurry. The cars all had palm-size Panic buttons installed on the outside, along with a remote-controlled switch designed so that judges or anyone else near a car that went off course could stop it. On the first day of the NQE, emergency stop, or E-stop, systems on each car would be tested, and any car that hesitated even the slightest bit in responding to the Panic button would be thrown out of the race.

During the race, each robot race car would be shadowed by a chase car, one of fifty Ford Tauruses, each of which had had its interior stripped of passenger seats and outfitted with roll bars and would be driven by a crash-helmeted professional driver. The drivers of those cars would have their own Kill switches to use if the cars they were following went astray. The so-called Kill switches actually had two modes—in addition to an out-and-out shutdown, equivalent to knocking a robot unconscious, the drivers and other race officials could also put the vehicles in pause mode, letting the car remain "awake," ready to roll back into action in an instant. As an added level of safety, the course was designed so that the three-foot-high concrete barriers, called K-rails, always separated spectators, judges, and other puny humans from the metal beasts.

Starting on day two of the NQE, and for the rest of the qualifying runs, the cars would have to demonstrate all of the maneuvers of the actual race, including turning into traffic, navigating from one GPS-defined point to another, and, in yet another sign that this was to be no ordinary auto race, finding and getting into parking spots.

I set up my laptop in the media tent, at the same table where Jan Walker would sit. For the next ten days, she and I would find the same fine red desert dust coating our laptop computer keyboards and screens, watch the race from the same video feed, and, I hoped, commiserate about the desiccating desert sun and generally become comrades in arms. Surely she would also see how hard I was working to bring to life every turn and slow-motion twist of the race for my readers, complete with photos uploaded to the Web as fast as I could snap them. I felt like a job applicant angling for a prized position at some elite organization where only the few, the proud, the brave need apply. Which, actually, wasn't too far from the truth.

THE IDEA OF AN AUTONOMOUS GROUND VEHICLE race had been kicking around DARPA headquarters since the directorship of Frank Fernandez, who had directly preceded Tony Tether, but it was Tether who finally codified it into the form it took in 2004. The wars in Iraq and Afghanistan provided a focus to the formerly vague idea of autonomous ground vehicles. Supply convoys winding their tortuous way through the battle zones were extremely vulnerable, a point that was brought home in the 2003 invasion of Iraq, during which a supply convoy was attacked and American soldiers killed and abducted.

Tether figured that a competition would be a good way to advance the technology quickly. And it was just a step beyond that decision for him to choose to bring in some maverick players to get fresh ideas beyond those advanced by the usual suspects. "He defined the idea of opening this up to the community," program manager Norm Whitaker later told me of Tether's approach for what became a series of three robot car races. The community was to include garage tinkerers in addition to the more traditional defense contractors. "I think that was the key insight that he had—that that would be of tremendous value," said Whitaker.

Norm Whitaker came to DARPA in 1998 fresh from AT&T's Bell Labs, where he'd worked as an electrical engineer specializing in optics. He'd been at Bell Labs since earning his Ph.D. from MIT in 1986, and he'd grown restless. "I had this really strong feeling," Whitaker told me later, "that people arrived to do pure research, to take their Ph.D. thesis and sort of repeat it and try to do it better. That was just wrong-headed." Instead of research for its own sake, Whitaker wanted to solve practical problems for customers. "I am much more passionate about figuring out what the right problem is that's going to make the biggest difference for the organization and working on that." The more he dug into his work, the more he found himself wanting to get the bigger picture. What kind of system would this bit of research I'm engaged in be useful for? Who would use this kind of system? Is this really what we should be building, or would some other configuration be more useful? Move far enough back from the problem, he discovered, and you find yourself out of the lab entirely. "I was thinking I really need to talk to customers: 'I wish I could hear what customers are saying because I'm not getting the right information from the lab.' "

At DARPA, says Whitaker, "Not only do you have to think what would be the technical feasibility of an idea, but you have to take it to the customers as well and understand how they do their jobs. You're talking to war fighters—people who actually use the systems—and figuring out what would truly make a difference for them." DARPA was heaven to him. "I used to tell people that the things I got in trouble for at Bell Labs were the things that I was rewarded for at DARPA," Whitaker told me. At Bell Labs, "I'd say 'We need to figure out who is going to be the customer for this,' and most people would say, 'Just be quiet. Let's just enjoy what we're doing here.' " At DARPA, "people want to see the entire package from end to end make sense before we even take the first step." His customer, the end user of the technology, was no less important than any other aspect. "DARPA is really a national treasure," Whitaker told me. "It's extremely unique in that way."

At DARPA Whitaker went to work on the Warfighter Visualization program, managing research into creating something like a fighter pilot's head-up display that would overlay a military commander's actual field of view with real-time computer-generated images showing the situational status of the soldiers under his command on the ground. He left the agency for three years and then returned as a contractor, ostensibly to work on projects in support of the war in Kosovo. But the first robot car race, 2004's Grand Challenge, had just ended, and Grand Challenge program manager Ron Kurjanowicz needed help setting up the next year's race. So Whitaker agreed to work on the project on a part-time basis. The assignment quickly ballooned into his full-time gig. "I ended up sort of pulling together the 2005 Grand Challenge as a contractor," he told me afterward.

His work on the 2005 Grand Challenge was Whitaker's first experience with robotics. To get up to speed, he picked the brains of everyone who had been involved in running the first Grand Challenge, to find out what had worked and what hadn't. Whitaker concluded that the event had mostly gone well, but that, in his words, "there was always little bits of chaos that could be improved." It turned out that running the Challenges was unlike anything else DARPA had done. Over the years, DARPA and other government agencies had developed a kind of shorthand with their contractors that smoothed the way for their working together. As Whitaker put it, "Things can be done relatively terse, and if you're not accustomed to it, it might seem like an unfriendly style." The trick to the Challenges was to find a way to do business that satisfied the requirements for the government's contracting system but that nevertheless included those garage inventors and other entrepreneurial types who ordinarily would never have had a chance to work with the government. Offering prize money instead of simply funding participants outright was the key to making it work—that and managing the motley crew of inventors, university researchers, and military contractors on their way to the starting line.

Maintaining a dialogue with the people who would eventually use the technology being developed was just as important as working effectively with those creating it. To help meet that challenge, program managers work with liaison officers—colonels and navy captains—within the different branches of the military. The liaison officers remain in contact with the DARPA program managers as they go about their own work, reporting back to DARPA when they find people within their own organizations who could use a technological boost. They also help the program managers understand how best to guide technology development so that it provides the greatest possible utility.

DARPA program managers keep their own eyes and ears open, too.

"I know people who have read articles in the newspaper," said Whitaker, "and just chased down the name in there of someone who sounded really coherent . . . and just got that person on the phone and talked to them." The point was to make that leap from experiment and lab work to practical use. "It's simply having lots of conversations, being willing to talk to anyone who is willing to listen to you," Whitaker told me.

The 2005 Grand Challenge wasn't long over before Norm Whitaker and his colleagues began putting together ideas for the next event. The robots had driven through the desert in the first two races. The five finishers in the last race had amply demonstrated their ability to navigate through open country and avoid fixed obstacles while navigating to GPS way points. The next logical step would be to up the ante with an urban-style competition complete with city streets and moving traffic. In scouting possible settings for the race, Whittaker and his colleagues found the Southern California Logistics Airport in Victorville, and that gave Whitaker what he needed to pitch the idea of the Urban Challenge to Tether. "When we found the site," Whitaker told me, "it helped to actually create what we were going to do. It's not obvious. . . . There were lots and lots of discussions [about] whether we were going to run it on a racetrack, whether it would just become a demolition derby. Would there be humans involved? Would we build a fleet of robots . . . that would become the traffic for the event? How would we test the vehicles?" The site itself provided the boundaries within which to answer those questions.

As for what Whitaker and Tether hoped to accomplish with the race, the way it would be perceived was as important as any mere practical accomplishments. Much of DARPA's work involves simply demonstrating that something can be done. Whether or not all the kinks in getting that something done are ironed out in the process is almost beside the point. The point is that innovators in a given field of endeavor see what is possible, and are inspired to push the envelope further themselves. "We always had this idea, you see, in order to expand the field, what we needed to do was create the visual of robots actually operating safely in traffic, and preferably with human drivers as well," Whitaker explained to me. He figured people wouldn't be that impressed by robots interacting with one another on those streets. But robots interacting successfully and safely with human drivers—now, that was another thing altogether. Before the Urban Challenge took place, said Whitaker, people he spoke to about it would laugh and say, "It's going to be a demolition derby." The goal was to remove the giggle factor. "We felt," said Whitaker, "that if people saw that [it] actually looked reasonable and robots could do this, it would change the way they thought about the entire field. . . . More than building a robot that was going to achieve any particular goal, the goal in this program was really the demo. The demonstration of what was possible."

In early 2006, Tether asked Whitaker to come on board as a government employee and DARPA program manager. Now it was time for the nitty-gritty. Determining the details of what the robots would actually be called upon to do in the Urban Challenge was no mean task. "We took the job of driving and broke it down into lots and lots of different pieces," said Whitaker, "and then took those pieces and decided some of them were too hard, or not necessarily too hard, but there were things we weren't going to test this time."

Determining whether a traffic light was red or green, for example, or detecting a stop sign, wasn't an important enough ability to test. A failure to perform those tasks would be hard to separate out from a more basic and critical function such as the ability to navigate through an intersection. Just perceiving the intersection, or knowing from an onboard GPS and map that it was there, should be enough, especially since future military planners would most likely want their autonomous supply vehicles to stop at all intersections, whether controlled or not.

No, the real challenge, Whitaker and his team decided, was navigation and obstacle avoidance, as in the two Grand Challenges, with the additional task of perceiving and reacting appropriately to other moving vehicles. This last task would be what really kicked the whole field of endeavor up to the next level. What's more, the event would act as a kind of practical robotics conference and get the best researchers in the field on the same page. After the event, those researchers should be able to collaborate or share information with one another much more easily.

The conflict between the need to clearly define operational requirements while leaving them vague enough for future applications created what Norm Whitaker called "a dynamic tension" between DARPA and the teams. "See," Whitaker explained to me, "the teams, if they can get you to define the problem well enough, then they'll build sort of a special-purpose machine that will do just that. If you tell them that every turn is going to be between eighty-five and ninety-five degrees, that's what you'll get, exactly that." To produce autonomous vehicles that would be truly useful to the military, the requirements for the robot race cars had to be less explicitly defined. "You want what appears to be intelligence in there," said Whitaker, "not just something that always just turns between eighty-five and ninety-five degrees." Giving the teams too little to go on, however, would create a problem that was too tough to solve in any reasonable amount of time.

I WANDERED PAST the bleachers set up next to the big tent at the starting/finish line, along a gently winding footpath through the dust that must have once harbored decorative plantings, past now-empty apartment-style barracks, past a tent where a vendor had set up shop selling sandwiches and drinks and snacks out of a silver trailer and where a pink metal pig blew smoke through its snout as it roasted meat inside, and to the team pits.

The teams had been installed in parking lots big enough for each to set up its own tent and trailer and still have adjacent parking areas and a workspace for their vehicles. The vehicles ran the gamut from friendly looking to downright scary—from the Golem Group's cheerful red Toyota Prius to Oshkosh's massive Medium Tactical Vehicle Replacement, or MTVR truck, called TerraMax. TerraMax was a Marine Corps vehicle manufactured by the team's sponsor, Oshkosh Truck, in Wisconsin, and was by far the biggest vehicle in the contest, although it was actually the smallest truck the company manufactured on its production lines. Its bigger siblings had six wheels instead of TerraMax's more conventional four. Even so, the beast weighed in at twelve tons and guzzled gas at a ravenous four miles per gallon.

Oshkosh had brought along some thirty people and a six-wheel MTVR variant, also called TerraMax, which had come in last among the five finishers in the 2005 Grand Challenge. To distinguish the vehicles, the team called the first TerraMax T1, a designation reminiscent of the killer robots in the *Terminator* movies. The team had also brought along a tow truck easily as big as the MTVRs. Oshkosh Truck was already supplying MTVRs to the Marine Corps, and the company looked at the Urban Challenge as a way to develop an autonomous version of the vehicle they already had in production. All of its sensors, from the laser range-finders mounted on the front bumper to the no fewer than eleven video cameras mounted at various points on the vehicle, were tucked out of sight, along with the GPS antennas and the computers running the monster. The idea was to allow marine commanders to combine human-driven vehicles with autonomous vehicles in their fleets so that casual observers wouldn't immediately detect a difference between the two.

The team to beat this year was the Stanford Racing Team, which had won the 2005 Grand Challenge. Walking past the team's pit area, I caught a glimpse of its twin blue Volkswagen Passats ensconced in their tent. Stanford had behind it all the resources of the university's renowned Artificial Intelligence Lab and sponsorship from Volkswagen of America, along with a cadre of roboticists and Ph.D. candidates at its disposal, including software lead Mike Montemerlo and team leader Sebastian Thrun, head of the Artificial Intelligence Lab. They didn't quite say that they thought this race was in the bag, but a certain aura of smugness betrayed their confidence. They had brought two cars to the race so that they could use one of them for spare parts should the need arise. Team sponsor Volkswagen had thoughtfully sent along a mechanic from company headquarters in Germany.

Like many of the teams at the Urban Challenge, Stanford preferred to use laser range-finders, called LIDARs (for Light Detection and Ranging), to paint a picture of the world around its robot, instead of more conventional video cameras. Video cameras are at the mercy of the sun and other external light sources that can change intensity and position depending on the time of day and their location relative to the car. Shadows that flitted and shape-shifted constantly would make an already challenging task even more difficult for a robot's computer brain.

Lasers, on the other hand, constantly beamed out invisible (to humans) but predictably intense beams of light that could cut through even the thickest shadows. Not that it mattered for this particular contest, but LIDARs did their jobs as well at night as during daylight hours.

The LIDAR units mounted on Stanford's Junior (and on other Urban Challenge vehicles) spun rapidly, throwing out their beams in all directions. The units picked up the return reflections and sent the resulting data to the car's software, which then was able to produce a view of the world made up of what Montemerlo called a "data cloud," a pointillist map of reflection points arranged in rings at incremental distances from the lasers' source, the car itself. By monitoring the ever-shifting rings, changing as the car moved relative to its environment, the software could pick out objects—for instance other vehicles—that moved at rates that differed from their background.

The other piece of the software puzzle was the planning and execution process. Once Junior's software had interpreted the changing environment around it, it had to decide how to act on that information. With route maps demarcated by GPS way points loaded before a run, Junior had to evaluate all the possible options for reaching its destination. The software ranked the various options in terms of risk. How much risk was involved in changing lanes, for example? That calculus would take in such factors as whether there were other cars in the lane Junior wanted to change into, whether those cars were moving quickly or slowly, and so on. One advantage Junior and the other cars on the course had over human drivers: their sensors could see in all directions simultaneously.

Where a human driver would have to crane her neck to see if anyone was approaching in that next lane, Junior would already know.

For the qualifying runs and the race itself, Junior's programmers would make sure that their machine drove with the proper degree of caution. Montemerlo admitted, though, that he and the team sometimes let off steam after the work day was done by running simulations with Junior's software aggression level cranked up into what they jokingly called Rambo mode.

At the other end of the spectrum of expense and expertise was Team Gray, named after the Gray Insurance Company of the New Orleans suburb of Metairie. Team leader and insurance company co-owner Eric Gray relaxed in a folding chair outside the team's two RVs as I toured the pits. The team's vehicle, a Ford Escape hybrid SUV dubbed Plan B, sat in front of the RVs, unsheltered by a tent. Gray explained to me that the car had gotten its name from the fact that it had had to be put together without any of the hoped-for DARPA seed funding that eleven of the other teams had gotten. Gray, a bearish silver-haired man in his fifties with an unpretentious, affable manner, spoke of his team's work as if it were the most ordinary thing in the world for an insurance man to become consumed with the idea of creating a car that could drive itself.

Sometime after the first DARPA Grand Challenge race in 2004, Gray read about the race in *Popular Science*, and got fired up about it. He dropped the magazine on his brother Michael's desk at the insurance company and got him interested in it, too. They decided to put together a team to compete in the 2005 race. The next obvious step (to them, anyway), was to discuss the idea with their company's IT department head, Paul Trepagnier.

The IT group was just coming off a project building a claims data and reporting system for use by the insurance company's clients, and they were going idle. Gray and his brother pitched the autonomous car idea to Trepagnier as a team-building exercise. It seemed like a no-brainer to them, but Trepagnier thought they were both nuts. "Impossible," he told them flat out when they suggested that he lead the effort to program a self-driving car. Just because he was the only one of them who could program computers didn't mean he could program a robot. He had absolutely no experience in that area. But he allowed himself to be talked into flying out to Anaheim, California, for a meeting of potential participants. By the time the trip was over, Trepagnier had found himself talked into competing as part of Team Gray, though he still thought that with no experience in robotics they wouldn't have a chance in competition against the likes of Stanford and other high-powered university teams.

Even so, in just the few months remaining before the competition, the team managed to successfully get its SUV outfitted with the computers and electronic sensors and controls it needed to drive autonomously. And then Hurricane Katrina devastated New Orleans. Most of the team's members lost their homes just five weeks before the race, and they christened their vehicle Kat-5, after the category-five strength of the hurricane. Astonishingly, even with their lack of prior experience and the disruption of the hurricane to their work, Team Gray ended the race as one of only five finishers, and only one of four to make it within the allotted ten hours. They had come up from nothing to the top five, in the same category as the likes of Stanford University and Carnegie Mellon. Not bad for a bunch of amateurs.

For this next race, Team Gray figured they were a shoo-in for getting the $1 million that DARPA was doling out to eleven of the competitors. No such luck. Gray surmised his team didn't get the money because no matter how hard he tried, he couldn't come up with more than $2 million in expenses for the project, so DARPA must have passed on him to fund teams with bigger expenses. The way Gray saw it, this was penalizing efficiency. "I guess I just don't know how the government works," he said to me.

So he went with Plan B to make the autonomous car project self-supporting: commercialize the team's autonomous car technology; form a company around it, called Gray Matter, Inc.; and try to sell it to auto manufacturers engaged in destructive testing of their vehicles—boring, repetitive tasks such as driving around and around a track until a car falls apart, or driving over extremely rough roads to test the limits of new vehicle designs. These aren't pleasant jobs for human drivers, but they're perfect for robots. Gray told me he was in talks with several major automakers, including Ford and BMW, and he had formed a partnership with a company making assistive devices to allow handicapped people to drive.

When the time came, I rode out with the team in Plan B for E-stop testing. We had hardly gotten out of the pits before the car started braking by itself. (The human driver was piloting with a remote controller about the size of a box of donuts in his lap.) The team drove the car manually back to the pit area, tore open the backseats, checked connections, ran diagnostics with a laptop, and ultimately traced the problem back to the remote controller. Luckily they had a backup control unit, from the 2005 vehicle. They swapped it in, and still made it to the test area in time for their scheduled test.

At their assigned E-stop test area, one of two side-by-side baseball field–size sections of desert demarcated by concrete K-rails, Trepagnier fussed with the computers in back and checked lines of code on his laptop before running the tests. The team members seemed just a bit rattled by the glitch on the way in and by the presence of the judges, but Plan B passed without another hiccup.

Next!

As a reminder that victory was anything but certain for the Stanford team, Junior showed that it was no more immune to the inevitable technical glitches inherent in perfecting new and sometimes quirky technology. During E-stop testing, Junior stopped cold before it even began, and refused to move forward. Team members hopped the K-rail barrier and headed out to their baby to see what was wrong. I watched them pile into the car and shut the doors like a troupe of circus clowns stuffing themselves into a toy car. All was quiet while the team worked code and ran diagnostics inside. Then, after a few minutes, both driver side doors flew open simultaneously and the engineers stepped out together. This happened enough times—Germans going out to run tests with the doors closed, doors popping open simultaneously as the men piled out of the car in unison—that I remarked on it out loud. "We are German," one of the team members standing on the spectator's side of the barrier with me said by way of explanation—only half jokingly, I thought. "We do things a certain way."

The problem turned out to be a glitch in the system the team had rigged up to read the controller area network, or CAN, data that the Passat, like all of its late-model brethren, generated about the car's status—things such as how fast it was going and which direction it was turning. Junior's computers tapped into that data to help guide the car, and without it, Junior was essentially blind. But for some reason, the computers weren't picking up the data, and until they did, they weren't going to let the car go anywhere. Good thing, too, otherwise the machine would have careened out of control, just as I later saw another vehicle do in more complex trials because of a similar glitch. Once the team diagnosed the problem, they simply rebooted the computer responsible for reading the CAN data, and the car was back in action, with only minutes to spare until the end of the team's time slot for E-stop testing.

Stanford wasn't the only team with a heavy German presence. Several of the teams, in fact, had mostly Germans on them. The only non-German on Team AnnieWay, for example, was a Chinese American named Annie Lien. She had been selected as team leader to fulfill DARPA's requirement that the teams be based in the United States and be headed by an American. Lien had had a hard time taking the job offer seriously at first, especially since her field was human-machine interaction, not robotics. She didn't dismiss the idea out of hand, however. The project intrigued her. "How much of my free time do I have to dedicate?" she asked Sören Kammel, the German computer scientist who first popped the question at the Palo Alto, California, research-and-development firm where they both worked.

"Maybe one or two days of paperwork, nothing more," he replied.

Sure, why not, she decided. By the time she found out that the administrative work of being the team's liaison with DARPA was far more involved, she was hooked.

I found her, a bemused and bespectacled young woman towered over by the Germans, in Team AnnieWay's area of the team pits. Her team-mates spoke rapid-fire technical German over her head, and she just smiled. She shook my hand warmly, clearly enjoying her role on the team, which by then the Germans had named after her. "We're the poorest team," Lien told me. They got only limited financial support from a Ger-man organization called the Collaborative Research Center on Cognitive Automobiles, which was administered by several German research labo-ratories and academic centers, so the team had to get by on a shoestring. Hence the play on words in its name AnnieWay—besides incorporating the name of the team's ostensible leader, it codified the team's commit-ment to winning a robotic race any way they could.

I asked Kammel why he thought there were so many Germans in the race. He smiled. "Germany is very proud of its automotive indus-try."

All thirty-five robot cars passed their E-stop tests, with Oshkosh's TerraMax putting on an especially exciting show. I watched as the mas-sive vehicle barreled along until a DARPA judge hit the Kill switch and the monster slammed to a halt amid a whoosh of air brakes and a cloud of desert dust. I think everyone watching was surprised at how quickly the big machine could stop, and the team was well pleased with their ve-hicle's performance. "Another box to check," said team leader John Beck with some satisfaction.

Now the teams moved on to the more demanding qualifying runs. Not all of their bots would survive.

"RIGHT NOW we're scared shitless," Sting Racing's software lead Mag-nus Egerstedt said as he watched with the rest of his team at the chain-link fence separating them from their baby, rolling away on its own for the first time. "We've never even done this before," said Egerstedt, looking for all the world like a nervous dad leaving his three-year-old at day care for the first time.

It was day three of the NQE. Here in Test Area C, the cars of the various teams had to stop at a four-way intersection in one of the dilapidated neighborhoods, figure out which cars had precedence at the intersection in order to decide when it was safe to move forward, and then navigate a simple round-the-block course. These tests marked the first time that the cars drove out of sight of their creators, and the members of the various teams craned their necks, anxiously watching for signs of their babies beeping and flashing along.

Following DARPA rules, the competing robot cars all had been outfitted with audible as well as visible indicators that were activated when the cars went autonomous. Just like the beeping backup indicators on construction vehicles and garbage trucks, the lights and sounds that were emitted tipped off bystanders that no one was watching where the vehicles were going.

There was no other traffic to interfere with Sting Racing's machine as it rolled up to the intersection. Still, it hesitated, feinted left, then began the right turn it was supposed to make.

"Whoa, that was excruciatingly scary." Egerstedt said as the bot completed its turn and disappeared from view.

The other team members cheered.

Sting Racing was a joint venture between the Georgia Institute of Technology, or Georgia Tech, and Science Applications International Corporation, or SAIC, a defense contractor. Egerstedt, a young Swede and a member of the faculty at Georgia Tech, downplayed his role on the project. "Mostly I just tell the postdocs what to do," he told me.

The team's senior faculty member was a Dane with close-cropped blond hair framing a round face. Henrik Christensen was officially known as "principal investigator" instead of team leader because of the DARPA rules stipulating that each team be led by an American. Christensen had little attention to spare for random questions from onlookers like me. He paced, gritted his teeth, and grimaced as word came back on what the robots were doing once they were out of sight. The DARPA race officials chatted away with one another on radios about what was going on, but everyone else, including team members, were on a need-to-know basis. As in "you don't need to know."

Sting Racing was upbeat about one thing, at least. "In terms of aesthetics, we've already won," Egerstedt couldn't help remarking to me. The team's Porsche Cayenne SUV was indeed a beautiful machine, painted deep blue and accented with a honeycomb pattern and the striped yellow jacket motif of Georgia Tech's mascot.

Egerstedt explained that the biggest challenge facing the team that day came from the fact that since their arrival in California they had been able to test their bot only in the open desert, instead of on the actual streets the car was now being called upon to operate on. Now, without the software tweaks needed to correctly calibrate the sensors whirling away on the car's roof and bumpers for the racecourse, it was having a hard time finding its way. Indeed, after it rolled out of sight of the spectators, it lost its lock on the GPS guiding it. Forced to operate on sensors alone, it lost track of where road ended and sidewalk began, and began climbing a curb before it realized it had gone astray. There it froze, trying unsuccessfully to regain its bearings.

A race official came to the chain-link gate to talk to the team. He gave them the option to pause the robot using the E-stop system and then unpause it to see if that would unstick it.

With the immediate difficulty resolved, the car again got rolling, and again the software interpreting its sensor data got confused. This time the car ran into trouble in full view in the four-way intersection on its way toward completing the test. Again, the car had lost GPS lock, and left to its own devices, it lost track of where it was. The stop line at the intersection was wider than the team planned for, and with GPS out of commission, the car didn't recognize the road ahead as part of an intersection. Without knowing it was facing an intersection, the car assumed the human-driven Ford Taurus DARPA cars waiting their turn to cross the intersection were simply parked, and it blithely rolled through the intersection without stopping to wait for the human-driven traffic to clear as it should have.

It was a despondent Sting Racing team that rolled out of the test area and back to their pit, looking at a long day of fixing code ahead of them.

Red Whittaker's demeanor later in the NQE trials was in sharp contrast to that of the members of Sting Racing. When I caught up with him at Test Area A, Whittaker—the head of the Field Robotics Center at Carnegie Mellon, and leader of the university's Tartan Racing team—was standing on a tree stump so he could see over the chain-link fence that separated him from the product of his toil of these last few months. Serene, his bald head framed against a milky early morning desert sky, he seemed almost a part of the landscape, and just as implacable.

The subject of Whittaker's intense gaze, a car named Boss, had stopped at the intersection on the other side of the fence and was awaiting its turn to cross. It was a blue-and-black Chevy Tahoe SUV emblazoned with a white numeral nineteen and logos for General Motors, Continental, Caterpillar, and Carnegie Mellon University, along with Google, Intel, and other lesser sponsors. A roof rack bristled with video cameras, laser range-finders, GPS, radio antennae, and a flashing yellow light that provided visual accompaniment to a steady beeping sound. Driving it was a stack of ten computer servers running the Ubuntu Linux open-source operating system and specialized software—some half a million lines of code.

Boss cleared the intersection and drove slowly out of sight behind an abandoned house. A couple of minutes later it reappeared a block away, still flashing and beeping, only to encounter the first of the obstacles DARPA officials had placed in the road to test the robot's ability to avoid collisions. It was a steel bar of the type used to block intersections at railroad crossings. The vehicle stopped before the barrier, then began a U-turn to head back the way it had come and find another route to its destination. Halfway through the turn, the car hesitated, LIDARs scanning the way ahead, processors evaluating distances and speeds to determine if there was enough space in the road to complete the turn without jumping the curb. While it considered, Boss let its front tires slowly move back and forth in a motion reminiscent of an insect tasting the air before it with its antennae.

A DARPA official with a walkie-talkie approached the gate to ask Tartan Racing if they wanted to put Boss into pause mode so that they could go out into the street and see what was ailing their robot. Whittaker remained aloof on his tree stump, his expression betraying none of his thoughts. And even as the DARPA man conferred with the team, Boss made its decision, completed its turn, and headed down the road.

I congratulated Whittaker on Boss's accomplishment. "Heads would have rolled if it didn't do that," Whittaker replied simply. "On a scale of one to ten, that was a four," he said. Despite his apparent detachment, the race was no mere exercise to Whittaker. At that moment, it was all about winning, and Boss reflected that attitude. "A machine like this is programmed never to quit," he said of Boss. "It wouldn't be okay if we didn't come out on top at every event."

Red Whittaker had found himself leaning toward robotics as a career choice as a young civil engineer in the early 1980s. "I was restless and technical and looking for something I could do with my own hands that would change the world and would fill in my time," Whittaker, a former marine, told me later. At the time, portable computers were just beginning to acquire the power needed to drive robots—that is, machines that could move about and perform tasks in response to programming or other input. "I saw a vision," said Whittaker, "for robots that would work in the world and farm and mine and secure the world and feed the world and explore the worlds beyond."

The years-long effort to clean up the radioactively contaminated Three Mile Island nuclear power plant near Harrisburg, Pennsylvania, gave Whittaker his chance to start testing his ideas in the real world. Whittaker refers to this time as the "primeval ooze" of robotics. "Robotics was certainly not a field of practice . . . it wasn't like there was a trade or jargon. There wasn't literature or institutions. There were no degrees in robotics." A perfect environment for a young and ambitious engineer to make his mark. After Whittaker's robots succeeded in reaching, and doing useful work in, areas too dangerous for humans at the power plant, his career in robotics was made.

Whittaker's group went on to build automated digging machines, volcano explorers, automated farm tractors, mine mappers . . . and race cars. "I heard about the first one," Whittaker told me of the Grand Challenge races, "and got cooking on the fourteenth of March back in 2003, knowing that I would be competing on the thirteenth of March in 2004." Carnegie Mellon's entry was a bright red 1986 Humvee called Sandstorm. "One of the things that really matters in a robot is having a great name," Whittaker told me later. "If you're going into a volcano, Dante is a good enough name, and if you're racing in the desert, Sandstorm does well enough." Sandstorm made it only 7 miles into the 142-mile Mojave Desert course. There it got stuck on a rock and, trying to free itself, spun its wheels until its tires caught fire. Thick gray smoke drifted across the dunes as the robot took its cue from its creator and refused to give up—no matter what. None of the fourteen other teams in the race got even that far. Half of them faltered before traveling even a mile.

The next year, Whittaker's team fared better, with its two entries—the original Sandstorm and a 1999 Hummer called H1ghlander (sic) taking second and third place to Stanford University's first in the 132-mile race through the Mojave Desert. "We had two great machines for [that] challenge," Whittaker told me, "two lead horses, untouchable. Just impeccable software. Thousands of miles of great testing. Performance I could dial in in a second." A kinked fuel line robbed H1lander of victory. "Boy, what a gut shot," Whittaker commented to me later, "when that fuel line pinched and that engine choked and it just didn't have the punch."

In 2007, Carnegie Mellon, with the backing of GM—Boss was nicknamed for GM's first chief of research and technology, Charles "Boss" Kettering—auto component supplier Continental, and a stable of other sponsors, was more than ready for the Urban Challenge. By then, Whittaker was dedicated, heart and soul, to competitive robot-building. "There is no question," he said to me, "that the challenge introduces a nonlinear rate of accomplishment that is unachievable in a mercenary context or something where you're paid to play." There was no better way to push the envelope in any technical field, as far as Whittaker was concerned, than through open, naked competition. "You can't buy that performance," he said. "And some of that is because you can't buy a human heart."

By the time of 2007's Urban Challenge, Whittaker had become more or less the grand old man of field robotics. He had graduated no fewer than twenty-five Ph.D. candidates in the field, and innumerable undergraduates and graduate students. Many of those former students had moved to other institutions, where some now worked on competing DARPA challenge teams—the Stanford University team's Mike Montemerlo, for example. Far from creating a rift between Whittaker and his former students, the races only served to strengthen their ties. "These are colleagues and associates forever," Whittaker told me. Watching his students get established in the world of robotics and make their own significant contributions gave Whittaker heart that the future of the field that he had helped to create was "in very good hands."

THE GOLEM GROUP'S cherry red Toyota Prius was in trouble. This machine was the minimalist of the competition. Instead of the racks of servers and extensive sensor arrays of the other cars on the course, the Prius had only one LIDAR on its roof with which to sense its environment and just a single laptop computer with a dual-core processor to gather data from that one sensor and make all the decisions the car needed to drive. One core, or central processing chip, interpreted the data coming in from the sensor using one software package, and the other core ran the decision-making and execution software. Now the car sat on the left-turn-and-merge course, called Test Area A, and refused to move forward.

The course tested the robot cars' ability to make left-hand turns in the face of rather heavy traffic created by the human drivers in their Ford

Tauruses. Making a left-hand turn can be dangerous, even for a flesh-and-blood driver, and the viewing area set up at this part of the Urban Challenge course quickly became the most popular spot for spectators because it was most conducive to accidents and because, unlike on the other two test courses, all of the action was in view.

The Golem Group's run had gone smoothly at first, with their car making seven or eight uneventful laps around the course, successfully stopping at intersections, waiting for the human-driven Ford Tauruses to clear, signaling, and turning left. But the car got hung up on a turn, with something in its thousands of lines of computer code telling it to stop, even though no cross-traffic confronted it. After giving the machine a decent interval to make up its mind, DARPA officials put it in pause mode and let the members of the Golem Group make adjustments. With tweaks made, the members of the team drove their car back to the starting point and bailed out, the judge blew an air horn, and the car again was left to its own devices.

This time the machine apparently felt even less like cooperating. It needed to straighten out its front wheels to head into the straightaway to reach the next intersection, the one directly before the spectators. As it turned the wheel to the left, it nudged the gas to drive forward. Problem was, the car's brain didn't get the message that the application of gas had been successful. So it gave itself more gas. And more and more. With the wheel still turned, it surged forward, hopped the curb separating the pretend roadway from the rest of the parking lot that contained the course, and kept bombing along, even after it blew a front tire. "Pause! Pause! Pause!" cried the judges. When the car finally shuddered to a halt, hubcap rolling away from the car like some kind of metallic tumbleweed, the machine was in no condition to continue the event. A flatbed tow truck soon pulled up to cart it away.

Red Whittaker's Boss seemed almost human as it gunned its way through the straightaways of Test Area A. It hesitated and backed up occasionally at the intersections, but it successfully waited for other cars to pass before executing confident turns. I watched it fly through twenty-six turns during its thirty- to forty-minute time slot. Finally, at twenty-eight minutes in, it hung up on the straightaway facing the spectators. After a restart, it headed toward the K-rail after its turn, backed up, slopped over into the oncoming lane, got beeped at by a human driver, and then kept going in the right lane. Twice it crossed perhaps a little too aggressively into oncoming traffic and got beeped at. Once, at a turn, it got confounded by a K-rail, backed up to clear it, got a beep from the car behind it, and continued on. Even with the beeps, which indicated potentially unsafe driving to the judges, this was the best show by far I'd seen in the difficult Test Area A.

A K-rail at the end of the straightaway facing the spectators' area presented the greatest challenge for the robots in Test Area A. It was perfectly positioned for witnessing the type of crash the Sting Racing team had suffered there. After heading down the straightaway and successfully turning through the intersection (left-hand turn signal flashing according to law, of course), the machine didn't turn sharply enough to complete the turn, and, whack, crunched the front right bumper against the K-rail separating the course from the spectators.

"This is awful," Egerstedt said, shaking his head back at the team's pit area. "We're down in the dumps," he told me, but "I'm ridiculously optimistic, so I think we're good." That was because "Remarkably enough, the sensors survived."

Despite Red Whittaker's intense commitment to winning the race, his Team Tartan did not hesitate to come to the aid of its rivals. At Tartan's invitation, Georgia Tech rolled their vehicle into Carnegie Mellon's well-outfitted shop tent and members of both teams went to work beating the bumper back into shape. When I expressed surprise to Red Whittaker later that such a competitive team as his would help a rival, he didn't blink. "It really is a community," he explained to me later. "It's a village. These competitions are really about a rising tide." A good competition, a truly challenging race helps all of the participants, Whittaker believes. And besides, his team members loved to solve problems. "Everybody around me lives for that kind of thing," Whittaker told me. "And I wouldn't have people who didn't love to do it."

After Carnegie Mellon finished the physical repair of Georgia Tech's car, the Sting Racing team stayed up late reworking their code. In the morning, they recalibrated their sensors and tested their software fixes in the designated team testing area. Thanks to Carnegie Mellon, they were back in the race, and hugely appreciative. At the start of the next day's runs in Test Area B, Sting came out of the chute displaying a big sign in the passenger window that read, "Thanks, Tartan!"

The chutes, defined by K-rails, were actually the starting gates that would be used by the teams on race day. The test called for the robots to leave the chutes, navigate through an open parking area, and find another narrow chute formed by K-rails to drive through a narrow road defined by more K-rails. Sting emerged nicely from its chute and headed for the exit, navigating with its GPS. But when it closed in on the K-rails that formed the boundary of the parking area, it hesitated. Idling, it considered the situation. It knew the exit was dead ahead, but it was having trouble seeing the opening it needed to get through. Finally, the bot started forward again. Then stopped. Backed up. Then idled, slowly oscillating its front wheels, pondering some more. Finally, moving very slowly, it found its way out of the parking area and through the chute, much to the great relief of onlooking Georgia Tech team members.

Sadly, the machine was totally stymied by a narrowing of the K-rails farther on. So the team went out in a DARPA SUV, tinkered, and then manually drove the car through the choke point. Back on track, the car moved along on its own until it came to a too-sharp curve—and promptly ran into a K-rail. Again, the car had crumpled its right front bumper. The damage to the vehicle itself was more extensive this time, but again the sensors seemed unharmed. Still, the accident killed the team's chances for qualifying runs for the remainder of the day—with its sensors again out of true, the machine would have to go back to the shop. And the team would again have to tweak their code.

DARPA officials had been somewhat overzealous in their use of those K-rail barriers. Such barriers had saved human onlookers from harm on more than one occasion during the previous Challenge races, and with this race even more demanding on the robots, Norm Whitaker and his colleagues were determined to take no chances. It was human safety before robot safety, and that's all there was to it. But the K-rails were too close together for most of the robots' comfort, particularly because DARPA's race rules stipulated that the cars keep a three-foot distance from the barriers.

For the biggest vehicle in the competition, Oshkosh Truck's TerraMax, that requirement turned out to be a physical impossibility. The machine snorted and grunted its way along at Test Area A, the left-turn-and-merge course, valiantly trying to stay within the bounds of the lanes while keeping that required one-meter distance from the K-rails bounding the course, but like an overweight airline passenger forced to squeeze into a seat built with smaller people in mind and encroaching on his neighbor's seat, the machine kept slopping over into the other lane.

Faced with the choice between disqualifying a well-performing machine and bending their own rules, DARPA officials finally moved the barriers on Test Area A to give the robots more room, letting all of the robots breathe a little easier. I watched TerraMax negotiate the newly resized course easily for eight laps without a hitch. DARPA drivers did honk at the vehicle three times, indicating that they felt the robot vehicle had cut them off. But from where I stood on the other side of the fence, it looked as if the machine had started its turns in time, but that its tremendous bulk could not clear the intersection quickly enough to suit the human drivers.

And so it went. The robot cars drove through the test runs in the three test areas, and over the course of the week showed DARPA judges whether they had the stuff to compete in the actual race. Mainly the judges were looking for vehicles that could drive safely enough to avoid taking out other cars on the open course. By the end of the National Qualifying Event, Tony Tether himself presided over the action in Test Area A, the left-turn-and-merge course, directing autonomous vehicles in and out of the course without regard to the test schedule. His goal was to narrow down the competitors as quickly as possible. Only twenty cars at most out of the thirty-five entrants would be allowed to run the actual race on November 3.

Team Case nearly eliminated its own robot, a custom-built robot car called Dexter, and the only one in the competition that wasn't a modified production vehicle. The machine had driven well in the 2005 Grand Challenge as the entry of Team ENSCO, managing to drive 81 of the 132 miles of the race before blowing a tire and dropping out. Here in Victorville, a team member, groggy from long hours spent on last-minute modifications to the vehicle, mistakenly plugged Dexter's laser range-finders into a power supply that delivered twice as much juice as they could handle. The forty-eight-volt power supply burned out the twenty-four-volt LIDARs in an instant.

With Team Tartan's earlier generosity still fresh in mind, Sting Racing, which had recently been cut from the competition, lent Dexter their own LIDARs. "Consider them organ donations," said Tucker Balch, Sting's team leader. Caltech ponied up the one additional LIDAR Dexter needed to get back into the race, and the repaired machine headed back into the qualifiers in Test Area B.

Dexter didn't do so well coming out of the chute, however, apparently losing track of where it was supposed to go and making a kamikaze run for the K-rails opposite the chutes. Officials paused it in time to avoid a crackup, and team members pushed Dexter back into position for another try, and again Dexter drove straight for the K-rails. Dexter's off-road tires grabbed the side of a K-rail and started climbing before one of the judges paused the machine. Clearly, Dexter had a bug that needed fixing, and the judges gave Team Case another chance later on. This time, Dexter found the exit chute and survived to drive again another day, though it didn't make the final cut. In fact, not twenty, but a mere eleven teams lined up in the starting chutes on race day. One of those was Virginia Tech's team, Victor Tango.

Like Team Gray in the 2005 Challenge, Victor Tango had come through tragedy to make it to the Urban Challenge. The number of Odin, the team's Ford Escape hybrid, thirty-two, represented the number of dead in a massacre at the team's home base. The previous April a crazed student had walked through a dorm and a classroom while shooting students and professors indiscriminately with handguns before killing himself. The brother of a Victor Tango team member had been one of the students wounded in the shooting, and many of the victims had been gathered in a classroom that the team used as its meeting space. "It's something we all had to live with and deal with," Victor Tango team leader and former Virginia Tech faculty member Charles Reinholtz told me of the tragedy's impact on his teammates.

Sharing the starting line with Victor Tango were Red Whittaker's Tartan Racing, Stanford, Team AnnieWay, Oshkosh, and six others. The rules for the race were straightforward. The cars would each have to navigate and maneuver through three "missions," which would be slightly different for each car. And they would have to obey the traffic laws of the local regime—in this case the government of the State of California—including a speed limit of thirty miles per hour. It was to be perhaps the world's strangest auto race. Not only would the cars have to conduct themselves as law-abiding citizens, but there would be no human beings behind the wheels, nor even directing the cars by remote control. These babies would be absolutely on their own. That is, except for the professional human drivers on the course with them—one for each of the cars, trailing at a discreet distance with hand poised over the emergency Kill switch in case any of the robots lost its printed-circuit-and-computer-code mind and went rogue—as well as thirty-nine other drivers to provide circulating traffic and vehicles with which to interact at the four-way stops.

The $3.5 million in prizes, plus the $11.0 million in seed money and the admittedly not inconsiderable cost of hosting this event, nevertheless added up to not a bad investment for a military agency seeking to develop the first cars capable of driving themselves through war zones. The bulk of the investment in these cars came from the automotive industry, universities around the country, and private companies hoping to score a piece of the action in what could well turn out to be the biggest

breakthroughs in automotive technologies since Charles "Boss" Kettering himself invented the electric starter at General Motors back at the
dawn of the auto industry. DARPA was yet again demonstrating how to
get the most bang for the military buck.

For anyone, myself included, who hoped to easily follow the progress
of the race—to know who was winning and who was losing—at any given
time, this course presented a challenging prospect. The cars would roll
out of the chutes at different times to give them a chance to get clear of
one another, and they'd each be driving different missions. They would,
however, return to the starting chutes after each mission, allowing me to
keep tabs at least on which cars were closing in on a final finish.

The DARPA press reps handed out a map of the course, with the
roadways across the dun-colored desert floor outlined in bright yellow.
The map showed a pair of wheel-like traffic circles fed by the spokes of
incoming roads. These would provide ample opportunity for traffic jams,
as Tony Tether and Norm Whitaker and company surely intended. Just to
provide a complete mix of driving conditions, one section went off-road
on the way to a straightaway. One team member assured me that this
would be the place to see, as he put it, some excellent "bot-on-bot action."
This section was perhaps most like a conventional race, though with that
thirty-mile-per-hour speed limit, it wasn't likely to present anything like
the pulse-pounding excitement of, say, the Indianapolis 500. Even more
bizarrely, compared to an ordinary auto race, the map outlined a red
parking area, actually Test Area A, where the robots would have to demonstrate their parking skills.

Victor Tango's Odin got out of its chute first, heading right for the
exit and out of sight on its way to the first of the traffic circles. The machine whizzed through the first of its three missions before the rest of
the bots were even out of their chutes.

Stanford's Junior; the Toyota Prius fielded by the University of
Pennsylvania and Lehigh University as the Ben Franklin Racing team's
entry; MIT; Team UCF, from the University of Central Florida; Cornell
University's Chevy Tahoe named Skynet (after the intelligent computer

system in the *Terminator* movies); and Oshkosh's TerraMax—all followed smoothly, one at a time.

Then the autonomous Ford F-250 pickup fielded by Team IVS got hung up heading out of its chute. It apparently had a crisis of confidence, making a U-turn to head back toward the chutes instead of toward the course's entrance. After the judges put the truck in pause mode, IVS team members manually backed it into its starting chute.

Meanwhile, Team AnnieWay launched its car, with Red Whittaker's Boss not far behind. True to form, Boss gunned its engine as it headed unerringly for the exit, as if determined to waste no time in accelerating to the thirty-mile-per-hour speed limit.

CarOLO, from one of the German teams, followed Boss, and by then Team IVS's truck was ready for another start. This time it got under way with no problems.

And so the race began.

Many of the teams didn't last the morning. AnnieWay's Volkswagen Passat, for instance, failed to enter a parking area properly and got pulled out of the race a mere twenty minutes after starting. The Subaru Outback belonging to UCF lost its bearings in the area formerly known as Test Area C, the four-way-stop and obstacle avoidance course, and pulled into the driveway of one of the abandoned houses, prematurely ending its run. Meanwhile, I'd gotten wind of a problem with Oshkosh Truck's TerraMax. I grabbed my camera and ran past the DARPA command center to the parking lot formerly known as Test Area A. There I encountered a scene reminiscent of a Japanese monster movie. It seemed the big yellow beast from Wisconsin had finally met its match. The machine had climbed the sidewalk in front of the former air force building the parking lot had been built to serve. There TerraMax sat, front bumper shoved against the building, which of course wasn't about to budge. After judges put TerraMax into pause mode, team members drove it to another parking lot not far away, where the race's losers were lining up. TerraMax joined Team AnnieWay's vehicle, already in the lot. By then only seven robot cars were still in the running.

Among the vehicles still in the race, those of Victor Tango, Stanford, and Tartan Racing were already distinguishing themselves as front runners. Victor Tango's whipped through its second mission and returned to the chutes before noon, and then, after only five minutes of prep time, headed back out on its last mission. Tartan Racing's and Stanford's vehicles rolled back into the starting lineup after completing their second missions just a little past noon, and headed out again in rapid succession. By mid-afternoon vehicles for only the three lead teams, along with those of Ben Franklin, Cornell, and MIT, remained on the course. The last two very nearly took each other out.

I looked up from my laptop in the media tent in time to see the immediate aftermath of the collision on the video feed. Cornell's ominous-looking black SUV, Skynet, had doored MIT's maroon Land Rover, called Talos II, in an intersection leading into a narrow channel formed by K-rails. I ran along the fence line to a spectator stand set up, conveniently enough, just a few hundred feet back from where the accident had occurred. "It looked like kind of a slow-motion train wreck," one spectator who had seen it told me. That was as good a description as any.

Skynet had gotten confused upon pulling up to the intersection. It knew this was where it should stop to watch for oncoming traffic before turning into the channel, but the GPS way point hadn't been centered properly on the intersection. It was off to one side, just a little too close to the K-rail barrier for the car's comfort. So it hesitated there, considering its next move. It tried backing up and moving into place atop the way point again, but still it was too close to the K-rail to make a confident approach. So it stopped again, again considering its options.

Meanwhile, MIT's vehicle, Talos II, had sidled up behind Skynet, heading for the way point as well. When Skynet didn't move forward, Talos II decided it was disabled and made a move to go around it. At precisely the moment Talos II was easing into the roadway ahead of Skynet, Skynet made up its mind to ease forward again, and . . . *contact*.

Race officials put the two cars into pause mode immediately. And now they sat, Talos II caught in mid-turn as it tried to cut in front of Skynet, both of them stuck partway out into the intersection. The cars hadn't been moving any more than five miles per hour, not enough to do each other any real damage. But they needed the help of their humans to disentangle them and set them on their way again.

Norm Whitaker was sympathetic to the bots' difficulties in that instance. "Humans wouldn't drive too aggressively there, either, I don't think," he told me after the race. It was one of those unexpected scenarios that might well have required human judgment to handle. "It's not a canned situation," he said. "In other words, they may have created canned routines for what to do in places like four-way intersections, but coming into this traffic circle area where there are sort of two lanes there, and one lane there, it's a little bit undefined."

Actually, Whitaker, in postrace analysis, thought the robots performed quite well overall. It was the human drivers who behaved unexpectedly. "We looked at the video afterward," Whitaker told me, "and you can see some of the vehicles on the course that stopped in the center of the lane, right on top of the stop line, and those are the robots." Robots behaving exactly as they were supposed to. Watching the tape, Whitaker and his colleagues saw other cars rolling through intersections with the barest pause that could hardly be called a real stop, driving the wrong way down supposedly one-way streets, and driving off the roads entirely, and those were all human drivers.

Whitaker had recruited many of DARPA's drivers for the Urban Challenge from an off-road racing organization called Score. These folks were used to a bit more excitement in their lives, not what turned out to be the tedium of the Urban Challenge. For instance, the endless looping around and around and around a parking lot in Test Area A. Eight to nine hours a day, driving ten miles an hour, honking a horn if a robot came too close to cutting them off. And wearing a helmet with the blazing sun of the desert beating down on the roof of their cars. "It was just brutal, a brutal assignment," Whitaker acknowledged of that particular bit of non-fun. Those drivers could be forgiven the need to occasionally cut loose or show a little more of their stuff during the actual race. It *was* a race, after all, and why should the robots have all the fun?

Tony Tether was on hand with a checkered flag to wave the cars in as they crossed the finish line. Three vehicles crossed ahead of the six-hour deadline. Stanford's cheerful blue VW Passat, Junior, came in first, with Team Tartan's Chevy Tahoe, Boss, not far behind, crossing at 1:45 in the afternoon. A couple of minutes later, Victor Tango's Ford Escape hybrid, Odin, crossed the finish line. Even though the order of the finishers didn't indicate which had won the first-, second-, and third-place prizes, it seemed clear that these three were indeed the winners.

Ben Franklin Racing Team's Prius, called Little Ben, crossed the line about an hour after the first three, at the same time that the final two robot cars, Cornell's Skynet and MIT's Talos II, were still rolling through their final missions. Talos II was having some difficulty on the off-road portion of the course. Like any good robot car, it had stopped to consider its options rather than barreling along on a potentially risky course of action. Eventually it found its way and rolled over the finish line forty-five minutes after Little Ben, followed only about a minute later by Skynet.

DARPA officials spent the remainder of the day and into the night tallying up score sheets. The next day, November 4, Tether took to a stage that had been set up on the finish line to award the three trophies, which took the form of small, medium, and large bald eagle statuettes. From Tether's point of view, getting rid of those trophies signified that the event had been an unqualified success. He beamed as he awarded the third-place trophy (and big display check for $500,000) to Victor Tango, second place to Stanford University, and first place to Tartan Racing.

Although the course's speed limit was thirty miles per hour, even aggressively programmed and clear winner Boss had averaged less than half that speed throughout the race, a pokey fourteen miles per hour. Junior's average speed was thirteen miles per hour, and Odin's even less. A small step for robot cars, perhaps, but a great stride forward for autonomous vehicle technology. The race, Tether proclaimed, was a "fantastic accomplishment." He said that the technologies developed for it were now ready to be developed for production vehicles and that DARPA could now step out of the picture. "DARPA is an interesting organization," Tether reflected. "We really never finish anything. All we really do is show that it can be done. We take the technical excuse off the table, to the point where other people can no longer say, 'Hey, this is a very interesting idea, but you know that you can't do it.' I think we're close to that point, that it's time for this technology to [be furthered] by somebody else."

What had started out in 2004 as a fringe event for garage tinkerers, with the odd university added to the mix, had transformed in three short years into a professional-level competition with sponsorship and technical input from major corporations at work on multimillion-dollar vehicles. Now the real work could begin. Although the problem of building cars that could drive themselves reasonably well under ordinary road conditions seemed well on its way to being solved, this still left the extremely tough challenge of getting a car to handle the inevitable unexpected situations humans encounter while driving.

Program manager Norm Whitaker described a scene not long after the Urban Challenge, when he was driving in to work at DARPA headquarters. A truck had stopped on the road ahead and the driver had gotten out and gone around to the back to secure his load, which had come loose.

Whitaker pulled up behind the half-dozen or so other cars that had stopped to wait for the truck driver to finish his task. "When I was sitting there watching that," Whitaker told me, "I was thinking 'What would my robot do?'" He doubted that an Urban Challenge–class robot car, even the best of them, could have analyzed the situation correctly. The car would have interpreted the scene as an accident, Whitaker thought, and tried to ease around the congestion before the humans in the road had cleared.

The appropriate action, which the human drivers took, was to wait until someone else got out of his vehicle to help the truck driver by directing traffic around him. With robots, said Whitaker, "we can drive eighty or ninety percent of the time and make the right decision. But the other ten percent of the time—how are we going to handle those cases?" Perhaps, Whitaker mused, it would be possible to build a PAL-type system that could learn just as humans often do, by observing other drivers. "Maybe you can have people drive vehicles and the vehicles sit there and observe driving situations and eventually learn to become a good driver by watching what people do."

But those were all just details to be worked out by others later. The point was there were people actively working on those problems now who hadn't been before. Take General Motors. Because of the race, GM CEO Rick Wagoner came to consider his company's involvement enough of a feather in his cap to brag about it afterward in his keynote address to the world's largest convention for the consumer electronics industry, the Consumer Electronics Show, in Las Vegas in January 2008. The company trucked Boss out to the show for display, and Wagoner told attendees, "Just to put this in perspective, autonomous driving means that, some-day, you could do your e-mail, eat breakfast, apply your makeup, read the newspaper, watch a video . . . all while commuting to work."

As a result of its participation in the Urban Challenge, GM became a convert to the whole concept of autonomous vehicles as the future of the automotive industry, and that, to Norm Whitaker, was one of the great successes of the event. Before the race, "they were sort of eating their own cooking all the time and they weren't branching out," Whitaker said of General Motors, "and just being out there with the university, they real-ized there was lots and lots of value in what was going on." Whitaker felt that participating in the robot car race gave one of the world's largest automakers a focus and a vision that it hadn't had before, "from the research lab, all the way to the top of the company." Whether or not that would help pull the company back to solvency in the midst of the world-wide economic storm that struck later in 2008 was another story.

I got a preview of the future of autonomous driving from a Continen-tal engineer named Michael Darms sent from the company's Germany headquarters to lend his expertise to building the autonomous driving systems aboard Tartan Racing's Boss. Continental, it turned out, was far from new to the field of autonomous driving, though Boss was the most ambitious system the company had yet worked on.

Darms was just a boy in the 1980s when autonomous vehicles made a splash in Germany with a demonstration of self-driving cars on the country's famous Autobahn. The cars were primitive by the new standard set by Boss and other DARPA Challenge vehicles, but they were enough to get Darms fired up in what became a lifelong passion. When he and I sat down at the Urban Challenge to talk about the future of autonomous cars, his eyes were bright with excitement. The first step for commercial-grade autonomous vehicle technology, he told me, was assistive driving features, some of which—such as so-called adaptive cruise control, which uses radar to keep tabs on cars on the road ahead and slow the car down or accelerate with the flow of surrounding traffic—had already been developed by his company and were available in some high-end production cars. New-generation vehicles also included systems to help drivers avoid accidents.

"Imagine you are driving down a highway," Darms told me, "and the kids scream in the background, and you look back, but the sensors continuously monitor the traffic in front of you. The car in front slams on the brakes. It's detected by the sensors, and you get a warning. You look forward again. And as soon as you touch the brake, the brakes are applied as hard [as necessary] to avoid the accident," or at least to lessen the impact. That technology, called extended brake-assist, has been on the market since 2005. Another system alerts a driver if she drifts out of her lane, for example by vibrating the steering wheel, to induce her to look forward. Darms and his colleagues at Continental figured a nonaudible cue, such as steering wheel vibration, was preferable to drivers because it would spare them the embarrassment of alerting passengers to their inattention. Besides, a vibrating steering wheel was similar to the effect of a car's tires passing over conventional rumble strips—already in common use to warn drivers of lane departure—and thus might garner a quicker reaction. Yet another system, called lane keeping, lets a car gently guide itself back into its lane.

"It's not autonomous driving yet," Darms said. "But with [the Urban Challenge] we can see that there is much more possible."

Current laws forbid automobiles from acting with true autonomy, but Darms was confident that in the foreseeable future cars would be given unfettered access to their own control systems. Darms, like GM's Wagoner, looked forward to the day when he would be able to focus on his e-mail or a newspaper during a road trip without the distraction of having to pay attention to the road.

At least one other professional on the course during the Urban Challenge was banking on a future of self-driving cars, but he wasn't a member of any of the race teams. Dean Collins was an analyst for Travelers, a major auto insurance company, and he'd come to scope out the future of his business. As we stood together at the fence at Test Area C watching the bots negotiate four-way stops with DARPA-driven chase cars, he filled me in on what he saw as the probable future of autonomous driving. He saw the Baby Boom generation as the early adopters of the technology, and he e-mailed me later to explain his reasoning. "According to the U.S. Census Bureau," he wrote, "the number of people over 80 years old in 2030 will be more than double that of the year 2000. The same is true of people over 65 years of age. These aging baby boomers are living longer and want to remain self-sufficient and mobile. We believe that this may drive a demand for robotically-driven cars, which could provide a safe, reliable alternative to driving oneself."

Baby Boomers, Collins and his colleagues figured, would reach old age at just about the time autonomous vehicles hit the market, ten or twenty years after the Urban Challenge, and that's when the technology would really take off. The Boomers would represent a big enough constituency to make sure the laws preventing fully autonomous cars were relaxed.

Well before then, the insurance industry would have to be up on all the ramifications of self-driven cars, chief among them, Collins believed, the question of liability. Liability was a fairly straightforward calculation for accidents involving conventional automobiles: generally, one of the drivers involved in a crash was at fault for that crash, and it was the insurance company of that driver that had to cough up for injuries or property damage. But who would be at fault in an accident involving autonomous vehicles, any one of which might legitimately have a human driver asleep at the wheel? Collins was willing to wager that it could turn out to be the auto manufacturers or their suppliers—whoever designed and built the systems that drove the cars. As Collins explained it to me, "Just as manufacturer defect can be a contributing factor in an accident today (faulty brakes, for example), this will continue to be the case as robotic technology evolves. In a future situation when the car is being driven robotically, and everyone in the car is a passenger, interesting questions about liability arise."

With any luck, accidents would be greatly reduced in a world of autonomous cars. Not only would the cars be more adept at spotting potential danger, with their 360-degree vision and lightning-fast reaction times, but they might also be in constant communication with one another—networked—so that each car on the road would know exactly what the other cars on the road intended to do at any given time.

CRAZY-ASS THINGS

DESPITE MY BEST EFFORTS, I had no luck breaking through Jan Walker's professional demeanor. She kept an intense focus on her work throughout the Urban Challenge, and though I greeted her and her staff each day with a sunny "Good morning," I never got more than a cursory response. Nor were my attempts to circumvent Walker by appealing to director Tony Tether and his program managers any more successful. Tether smiled when I hustled to the front of the press tent after a press conference to shake his hand and introduce myself, and kept walking. "Go through Jan," he told me, when I tried to speak with him. "I do whatever she tells me."

After the race, I approached Urban Challenge program manager Norm Whitaker, who was a bit more talkative. "I guess Jan told you about the book I'm writing on DARPA?" I asked him.

"No," he said, but he seemed interested.

The Urban Challenge put me in tantalizing proximity to DARPA program managers, who were acting as judges for the event, and to the director himself, but without Walker's say-so, I couldn't interview them for the record.

My experience at Urban Challenge was a replay of the one I had had at DARPA's trade show, just two months earlier, and from which Jan Walker had been absent. There, too, I'd been invited to a DARPA function crawling with program managers without real access to the agency itself for interviews. The DARPA Technology Symposium, or DARPATech, is held at irregular intervals—the next one would take place in 2010—and brings together DARPA program managers, military brass, contractors, and contractor wannabes under one hotel roof so that DARPA can show off its progress on unclassified programs, hear about ideas for new programs from the field, and recruit new program managers and contractors. The result is something of a cross between a science fiction convention and a job fair, but far more interesting than any science fiction convention I've ever attended.

The 2007 DARPATech was held in August at the Marriott Anaheim, in California, within easy walking distance of Disneyland—which seemed appropriate. "I think being at DARPA is like going on a trip to Disneyland," said David Lucia, a program manager at the Tactical Technology Office in a recruitment video on the DARPA Web site. "You know," he elaborated, "time there is short, and you just run from ride to ride, and you try and get in as much as you can because you know the day is going to end, and you're going to go home, and life's going to go back to normal, and your parents aren't going to get up the gumption to go back anytime soon. Being at DARPA is like that. You know that your time is short. You're at an amazing organization. You have a very limited time. You want to work on all the program activities, all the things you can do that you have an opportunity to do, and so you're running around here, you're thinking, you're working, you're traveling fifty times a year, and you're exhausted, but you're having a great time. And at the end of it, hopefully you look back and you say, 'Wow, maybe, just maybe, I did something great.'"

DARPA managers are constantly on the lookout for new researchers and their ideas—as well as their own replacements. Since they serve only four- to six-year terms, the agency is always on a recruiting jag, and it devotes significant resources to the activity, including the three-day event, which this time attracted some three thousand attendees, including army, navy, and air force brass; (mostly civilian) program managers; and a select group of researchers and others who wanted to work for DARPA. Program managers delivered presentations on their work, complete with flashy videos, and contractors conducted hands-on (literally, in the case of the prosthetics program) demonstrations of applied science fiction.

DARPA's need to recruit ultimately created the crack in the door I needed. The agency is so desperate in its search for new program managers that a program manager actually tried to recruit one of my editors at Wired.com, Noah Shachtman, who writes a couple of thousand words a day on military technology for the site's *Danger Room* blog. Shachtman told me that he refused to consider the idea seriously.

Along with the presentations and the demos, a third major focus of the 2007 DARPATech was a series of so-called sidebar discussions, where researchers and technologists of all stripes could pitch their ideas directly to DARPA program managers. The sidebars took place in a vast exhibit hall filled with cubicles with doors that could be shut for privacy. Anyone attending DARPATech could register for one of the fifteen-minute sidebars, though, as I found out when I showed up for my session, anyone with a press badge would be turned away at the door. So I went to the long line of people waiting to get in for their sessions and talked to a slightly rotund man in glasses and a quietly intense middle-aged woman standing with him. They were pleasant enough, and they told me they were in materials science, but they wouldn't tell me their names or the name of their company, and they certainly wouldn't tell me anything about what they were pitching. Even so, the presentations and the displays of the various DARPA program offices gave me plenty to keep me occupied.

DARPA's six program offices (two of them merged in 2008, leaving just five) were charged with funding research on particular areas of DARPA hard technology, and each of them had a display booth in the main exhibit hall. The Defense Sciences Office, or DSO, oversaw such "far side of the far side" projects as attempts to describe mathematically the workings of complex organic systems—such as, say, the human brain—with a view toward duplicating their functions in artificial systems. It was in the DSO booth that I caught up with Jon Kuniholm and the other members of the Revolutionizing Prosthetics program just a week after my visit to their headquarters at Johns Hopkins University.

The Information Processing Techniques Office, or IPTO, that had funded the ARPANET and other research at SRI continued to work on extremely challenging computing projects, such as an attempt to develop handheld real-time translation systems so that people speaking foreign languages to each other could make themselves understood.

At one IPTO display, researchers from the University of Southern California stood with a researcher from SRI. At their side-by-side DARPATech displays, they demonstrated two speech-translation systems running on laptops. The USC system translated between Farsi and English, while the SRI system translated between Iraqi Arabic and English. USC's system allowed me to speak into a handheld microphone. On the first try, the computer accurately interpreted my spoken English; I watched as my words appeared on the screen as quickly as I uttered them.

I had had some experience with commercially available speech-recognition technology that required me to specially train the software to recognize my unique speech patterns. Afterward, it would recognize only my speech, and then only in a quiet room. The IPTO system took speech recognition to a whole other level. Here I was, a first-time user of USC's system, speaking in a crowed conference hall full of ambient noise for the system to sort through, including competing voices from people standing nearby, and the system got it right the first time, in real time. I was impressed.

What happened next, though, gave me goose bumps. Only a second or two later, the computer spoke my words back to me in Farsi. The Ph.D. student standing beside me listened to the computer's spoken language, replied in Farsi, and, after that brief lag, the computer replied to me in English.

The system had to perform several distinct tasks, not the least of which was accurately recognizing my speech in much less than ideal conditions. Next the system had to translate my speech into another language, and, finally, repeat those words as spoken language. All of which the system was apparently performing admirably well.

My mind whirled with the possibilities. Here was a device that brought people of different cultures together smoothly and efficiently. These translation systems, once shrunk down to handheld devices, were designed to enable American soldiers deployed in foreign countries to communicate more easily, but perhaps such devices would remove at least some of the impetus for conflict in the first place.

People don't always trust their human interpreters in the field, Kristin Precoda, the director of SRI's Speech Technology and Research Laboratory, told me. Human translators have their own motivations, fears, and alliances, which put their objectivity in question. At times there haven't been enough of them to go around in Iraq and Afghanistan anyway, and the few who were available were often at risk. This way the human translators could be reserved for the most sensitive and complex tasks.

The Microsystems Technology Office (MTO) concerned itself with the technology of the very small, including microchip-size atomic clocks and quantum computing, an emerging field that promised to vastly increase the power of computer processors. A notably startling MTO project involved implanting electrodes and computer chips in insects to see if they could be remotely controlled. "Using insect-integrated electronics, we might be able to control insect flight," program manager Amit Lal announced to the audience of his DARPATech presentation. "Much in the way we have farmed horses and camels as beasts of burden, we might now consider farming insects for robotic missions." Lal and his colleagues envisioned implanting insects with tiny cameras and other sensors and flying them into danger zones ahead of soldiers.

The Strategic Technology Office (STO) funded technologies that could influence the overall context of a conflict and that spanned more than one of the branches of the armed services. Systems for detecting and knocking out hidden enemies and their weapons fell under its purview, along with a group of programs that could produce some of DARPA's most important work yet: those for enabling armed forces to generate their own power in whatever environment they found themselves in. The goal of program manager Douglas Kirkpatrick was nothing less than setting the armed forces free of what he termed the "tyranny of oil." Programs under Kirkpatrick's purview aimed to radically increase the efficiency of standard solar panels and to create high-performance fuels from vegetable oil rather than petroleum. The potential benefits for the military were obvious. "Throughout history, energy has been *the* limiting factor in all military operations," Kirkpatrick said in his DARPATech presentation. But the civilian world stood to gain equally as much.

The Tactical Technology Office, or TTO, fostered development of "high-risk, high-payoff" technologies for fighting battles, including offensive weapons. In that sense, the office represented perhaps the purest embodiment of DARPA's original raison d'être as envisioned by Eisenhower's secretary of defense McElroy and his scheme for developing the "vast weapon systems of the future." TTO was helping DARPA get back into space, with, among other projects, robot satellites that could refuel and upgrade other satellites in orbit. On display in the TTO booth was a Merlin rocket engine—taller than a man, with gleaming chrome tubes and a flaring exhaust nozzle—built by Space Exploration Technologies and used in DARPA's quick-launching rocket program. Less flashy, but perhaps even more intriguing, was a scale model of a black, rapier-prowed airplane powered by proposed hypersonic—that is, capable of flight five times or more above the speed of sound—air-breathing engines.

Presentations in the big auditorium by DARPA program managers from each program office were emceed by DSO program manager Amy Kruse, a thirtysomething neuroscientist and a seemingly endless fount of energy. I was particularly struck by the presentations of the Defense Sciences Office. DSO program manager and former Harvard University professor Benjamin Mann and his researchers sought answers to advanced problems in algorithmic origami, network dynamics, and many others. They also wondered whether it would be possible to create a kind of biological quantum mechanics theory. Such a unified field theory for biology, as Mann termed it, would enable scientists and engineers to predict and even duplicate the behaviors of complex, and as-yet-unfathomable, biological systems. Mann and his colleagues sought to reduce all the complexity of life to simple ones and zeros, and thereby gain mastery over them. One application, if they were successful, would be the ability to quantify, predict, and ultimately even duplicate the inner workings of the human mind.

Other DSO programs presented at the conference included a project to bring the cost of titanium down from its present prohibitive $35.00 per pound to a bargain basement $3.50 per pound. Walking autonomous robots funded by DSO were being built to carry supplies over rough terrain for soldiers, but could prove useful in civilian applications as well. Technologies allowing people to adapt to extremes in temperature or high altitudes; or to dive underwater while using 30 to 40 percent less oxygen, just as dolphins do; programmable materials that could form and reform themselves into different useful shapes on command; growing plants that respired hydrogen to be harvested as a cheap source fuel—all were being developed by DARPA's Defense Sciences Office.

While DARPATech, and a short time later, the Urban Challenge, gave me a window into the action at DARPA, I still needed deeper access. I was beginning to feel desperate in October 2007 when I was approached by TV producers Steve and Ellen Eder, who ran a production company called Terra Nova Television. The Eders wanted to create a TV show about private space travel, the subject of my previous book, and they came to me for advice and direction. When that idea didn't fly among the TV networks the Eders pitched it to, I suggested they do a show about DARPA, with my help.

That idea stuck, eventually landing at the National Geographic Channel, and I gained a pair of valuable allies in getting in the door at DARPA. The Eders and I applied pressure to Jan Walker together, and finally, she cracked, granting me an interview with DARPA director Tony Tether himself in January 2008. This was the break I needed. If, in a face-to-face meeting, I could pitch myself to Tether as one-stop shopping for

both a TV show and a book about DARPA that could help in his recruitment efforts, I figured I had a real shot at being granted the access the
Eders and I both needed to complete our projects.

DARPA today occupies a ten-story red-stone-and-smoked-glass
building in Arlington, Virginia. Nothing distinguishes 3701 North
Fairfax Drive from any of the other unremarkable office buildings in the
neighborhood, not even a sign or a logo. There's a baseball field behind it,
a bank branch across the street. The building entrance sits right on the
corner, two step-tiered wings flanking a set of white cement steps leading
up to the entrance. A flagpole flies the Stars and Stripes in front. Only a
lone policeman walking back and forth in front of the entrance, a black
unmarked Crown Victoria idling in an illegal parking zone just down
from the corner on Fairfax, hint that this is no ordinary office building.
The lobby inside is similarly plain. In fact, the place is notable for its lack
of distinguishing features, except for the pervasive scent of freshly fried
fast food emanating from the Festival Café just within. On any give day
at lunch hour, a visitor is likely to see DARPA program managers in suits
and ties picking up sandwiches in the café to bring back upstairs to their
offices, along with the occasional air force, army, navy, or marine officers
working with the program managers.

Around the corner from the main entrance, past the elevator banks,
where a single security guard armed with an automatic sidearm checks
the badges of employees, a couple of not particularly intimidating security guards man a security desk with the DARPA logo hanging behind
it. They're friendly, even to the point of cracking jokes about DARPA's
shadowy presence in popular culture—video games with plots about
kidnapping the director of DARPA, tunnels in the basement leading to
secret government installations, and the like.

The office building is rented, not government owned, and I got the
distinct impression of *temporariness* on my first visit as I exited the elevator
on the ninth floor, home of the agency's executive offices, with my escort,
one of the guards from the desk. As on every floor, the elevator landing
is sealed from the offices on the rest of the floor by heavy, windowless

doors festooned with "Restricted Area" warning labels and admonitions to "Ensure this door is secure behind you!" Jan Walker greeted me at one of the doors, and the guard from downstairs handed me off to her.

The walls beyond the door were all white and, like the lobby downstairs, devoid of decoration, except for a row of presidential portraits arrayed in chronological order above a row of settees. The offices on this floor all had unmarked wooden doors, some decorated, dormitory fashion, with military posters or other temporary-seeming decorations. I got the feeling that the occupants could have vacated the place within hours, at a moment's notice, with nothing but a few stray papers fluttering to the carpet. "Move along," the cop out front might say, "nothing to see here, never was."

The unmarked door across from the settees just off the elevators turned out to belong to DARPA director Tony Tether himself. Tether, a round, balding fellow in his sixties with thick glasses, dressed in a shirt and tie, is the sort of person who would draw absolutely no notice whatsoever waiting on a corner for a bus at the start of rush hour in any major U.S. city. He looked a bit surprised when he opened the door and saw me there with Walker. "I thought this was going to be a phone interview," he said. I smiled and shrugged, as if to say, "I happened to be in the neighborhood, so it was just as easy to drop by as to phone." Actually, I had just driven six hours from my home in New York State.

Walker and I settled on a couch in Tether's office facing his big hardwood desk, while he sat in a chair on our side of the desk and folded his hands over his wide belly. He regarded me speculatively.

With my digital recorder off, I gave him the pitch. First I handed him a copy of my rocket book and a magazine for which I had written a cover story about one of his projects. He nodded approvingly at the latter, and I told him to call some of the people I had written about in the former as references for me. I filled him in on the TV production and my role as advisor to it, and on the book project, which I pointed out would be the first of its kind. My goal, I told him, was to bring the agency's accomplishments wider recognition. He immediately saw the potential value of both projects as recruiting tools. He warned me that I had better portray his agency accurately, or I'd never be allowed in the door again, and then, just like that, he granted me access. I felt like a program manager pitching a new idea and getting an instant yes. I tried to keep the grin off my face as I reached over to turn on my recorder.

There were restrictions, of course. All of my interviews with program managers would take place with Walker present or listening in and making her own recording, just as in that first interview with Tether. Interviews with program managers could take place only in a second-floor conference room, with no opportunity for me to explore the rest of the building (with one notable exception) or to sit on meetings and otherwise observe the daily workings of the agency firsthand. And I'd get to learn about only the portion of the agency's work that was not classified. But I, and through me the Terra Nova film crew, now had unprecedented access to the agency. At last I could tell the story of the agency through the people who worked there and get behind-the-scenes details on the projects I wanted to cover.

It didn't take long to get Tether talking about his job. "What excites me personally about my work is that I have been here six and a half years and I have never left here any day without learning something new. I mean really new, too. I mean, six and a half years. Two thousand days, and something—at least one thing—brand new. Plus the fact that we're doing great stuff. On a broad front, from our science office to our big systems offices, we are doing things that are just incredible—most of which you will never hear about."

"DARPA's original goals were to prevent technological surprise," as Tether reminded me during a later conversation. But "over the years we have found that the best way to prevent technological surprise is to create it." Hence the need for secrecy. "If you're going to create surprise," Tether pointed out, "you can't tell anybody what you're doing—otherwise it won't be a surprise." He cited stealth aircraft as a top-secret technology created by DARPA that gave the U.S. Air Force an edge no other nation on Earth had, or even knew about initially. Stealth aircraft are an example of what Tether and others refer to as game changers. "After World War Two everyone was putting up radars to detect incoming enemy aircraft. And radars worked very well. But once we took that away from them by creating stealth aircraft, we changed the whole game."

Changing the game first has been an important mission for the agency from the beginning, and while I wouldn't be given even a peek at secret game-changing projects in progress, stealth technology offers an insight into the kinds of technologies that might be in the works today. Stealth—or low-observability, as it is more prosaically known in defense circles—was a carefully guarded secret from its beginnings as a DARPA program in 1975 until the first stealth airplane was finally revealed to the public through a deliberately blurry photo in 1988, fully seven years after the first prototypes had been built.

Lockheed Martin's famously secretive Skunk Works division rose to the DARPA challenge to reduce the radar cross-section of airplanes, cut the amount of heat coming from their engine exhausts, and otherwise reduce their detectable emissions to such an extent as to make them practically invisible to enemy defense systems. The resulting prototypes, called Have Blue, featured a now-famous multifaceted aircraft design whose flat planes and sharp angles made it look as though it had been designed by Pablo Picasso. The strange geometry, antithetical to an aerodynamicist's idea of good form, was an artifact of the limitations of the computer program the engineers used to calculate the optimal radar cross-sections for the aircraft. The specially developed Echo 1 program, and the computers it ran on, lacked the power to calculate radar-deflecting shapes in anything but two dimensions, and so its solutions all took the form of flat planes, which engineers combined to create their aircraft. Thus designed for stealth over aerodynamics, Have Blue was inherently unstable in the air and depended on computerized flight controls to stay aloft. Nevertheless, flight testing beginning in 1977 proved that the design rendered the airplane extremely hard to detect by radar, and nearly impossible to lock on to with weapons systems, making it ideal for slipping into enemy territory to knock out radar and missile installations in advance of bigger bombers and fighters.

Lockheed followed up Have Blue with the F-117 Nighthawk stealth ground-attack aircraft for the air force, which flew fifty-nine of them through the 1980s until 2008, when the advent of the stealthy F-22 Raptor made them obsolete. Although no longer forced to rely on the Cubist flat-plane design, the new stealth aircraft nevertheless owes much to the technology developed with DARPA funding for its predecessors. The B-2 stealth bomber, still in operation at this writing, also owes much to Have Blue.

DARPA's greatest asset, Tether told me, is its program managers, and greatly contributing to their success is their limited tenure at the agency. "The term limits are what really make this place," said Tether. "When someone walks in the door here, there's a sense of urgency that you don't find anywhere else in government."

In contrast to employees of other government agencies, the program managers—physicists, chemists, engineers, mathematicians, and others with crazy ideas to farm out—all serve terms of limited duration. The agency doesn't discriminate against foreign researchers—good ideas are good ideas, wherever they happen to be hatched. The program managers themselves, however, have to be U.S. citizens. On one of my visits to DARPA headquarters, I met a former DARPA researcher from the United Kingdom who was so taken with DARPA while doing DARPA work as part of his duties as an aeronautical engineer in the British Royal Navy, that he became a U.S. citizen in order to join DARPA as a program manager. His transformation to an American was complete, he told me, except for his British accent, which he seemed genuinely to regret not being able to eradicate. The one time that I was able to leave the conference room where interviews were conducted, albeit sans my notepad and recorder, I followed him back to his office. An American flag stood in a corner, and shelves on his wall held, in addition to scale models of military aircraft, portraits of his daughters, who were now serving in the U.S. Army. The man's commitment to DARPA was especially striking, given his limited tenure there.

"A typical DARPA term is three to four years," program manager

Douglas Kirkpatrick explained to me, "and a typical maximum DARPA term is six years." There's no danger of forgetting; every program manager has an expiration date printed on his or her badge. "If there was any word I would [use to] describe that is uniform across all DARPA program managers," said Kirkpatrick, "it's just how intellectually aggressive we are. It is in part due to the fact that all of us come in seeing the clock ticking . . . and get the hell out of our way because we've got stuff to get done. That's what we're about. We're rather impatient people, and in the government, that's a good trait."

That "stuff" to get done is usually something a program manager had been working on before coming to DARPA. "The people who come here have an idea that they just can't get done where they are," Tether told me. "They have a feeling that it can be done—not necessarily the knowledge of the exact details . . . but they have a feeling that it can be done, and if it can be done, that it's going to be wonderful." Something wonderful such as radar-invisible airplanes, titanium as cheap as aluminum, cars that drive themselves through traffic, artificial limbs indistinguishable in form and function to a casual observer from natural ones, or any number of seemingly impossible ideas. Incoming program managers "know that they're not going to get their idea finished in the four years that they're here," Tether said. But "they know that they can actually get it started and get it through those first early hurdles." They know, in other words, that they might be able to make the impossible possible. It's an irresistible challenge to the right kind of mind.

That drive to pull off the impossible is what motivates the program managers more than anything else. They're certainly not driven by money—not with a government salary. They don't get stock options and, working in obscurity on often-secret projects, achieve little renown on which to capitalize for their next gigs. In fact, says Tether, "with some of the ethics laws, it's sometimes unclear whether they can even get a job—other than maybe getting a franchise in a donut shop." The minimal financial incentives do complicate recruiting efforts, but they also help the agency attract talent motivated by passion, not greed. "DARPA program managers never take no for an answer," said Tether. "I can say, 'No, no, no,' and they will come back, dress it up differently, sometimes not even bother, and still keep coming back to get their ideas done."

As in the old days, the development of a program still comes down to a pitch meeting between the director of DARPA and a program manager who wants to start a pet project. The kind of quick pitch meeting that resulted in the funding of the computer network that became the Internet was the rule, rather than the exception, then and now. With just two layers of management—a program office director and the head of DARPA himself—program mangers soon get used to getting a maximum amount of exciting work done with a minimum of paperwork. This often creates a rather heady atmosphere around the office, where a program manager can dream up a wild-haired idea in the morning, work up a pitch hunched over a sandwich at his or her desk over the lunch hour, and then get the idea funded as a DARPA program in the afternoon. To be sure, the lack of paperwork has its downside. Records of past achievements are spotty at best, and the term limits result in a sketchy institutional memory as well. The agency looks ahead, not back, thanks in large part to a 25 percent changeover rate in its management staff every single year.

What does it take to become a good program manager at DARPA? "The best DARPA program managers, I swear, are science fiction writers," Tether told me. To DARPA, ideas are everything. Tether considers the execution of them—at universities and private companies around the country—almost an afterthought, because, as he put it, "There are thousands of people we can go hire to execute an idea. But it's coming up with it in the first place. . . ." He cited as an example the famous science fiction writer H. G. Wells, who coined the term *atomic bomb* in his 1914 book *The World Set Free*, describing for the first time a weapon harnessing the only recently discovered power of the atom. "Now, he had not the slightest idea of how the hell to do it," Tether said, "but he knew enough that, wow, this would be a great capacity, a very small package, have tremendous energy."

With Wells thus recast as a DARPA program manager, the scientists of the Manhattan Project, who built the first nuclear weapons at the end of World War II, would be merely "performers," in DARPA parlance, bringing to fruition the more significant ideas of Wells. Taken to its logical conclusion, Wells should get credit (or blame) for the creation of nuclear weapons, not Oppenheimer, Teller, and the other scientists who laid the theoretical groundwork that made nuclear weapons possible. Of course, that's not the case, because they did the actual work of making it happen, not Wells. Interestingly, the opposite is true of another modern development: communications satellites. In a paper published back in 1945, science fiction author Arthur C. Clarke outlined the principles by which an orbiting transmitter could maintain contact with a given area on the ground even as the planet turned beneath it. That was almost two decades before the feat was actually accomplished. Today Clarke gets the credit for the invention, even though he built none of the hardware for that first geosynchronous communications satellite. In fact, the precise position in space that makes communication satellites possible is often called the Clarke orbit.

Fair or not, Tether's statement explained an institutional attitude on DARPA's part that I had until then found puzzling. Namely that the agency almost never gave credit to its so-called performers, choosing instead to take full credit for all of its programs on its Web site and recruiting materials. To find out who was turning those programs into working hardware and software took a bit of detective work on my part, searching backward from the name of the program to find the institutions, companies, and inventors actually putting bolts to metal, soldering circuits, writing code, and all the thousands of other tasks required to build the bionic arms, aircraft, rockets, and robots that DARPA's program managers envisioned.

Throughout its history, DARPA program managers have prided themselves in remaining open-minded about ideas that might at first glance appear outlandish or even absurd, as long as there was some chance that they might give new advantages to the nation's armed services or that an adversary might realize them first. Joe Mangano, who was deputy director of DARPA's Directed Energy Office (i.e., office of ray guns) from 1978 to 1984, and who returned to DARPA in the 2000s as a program manager in the Microsystems Technology Office, put it this way: "Someone who says something can't be done and can't prove it is someone to be avoided by DARPA. You've got to fail some fraction of the time or you're not pushing the limits far enough. And DARPA, unlike most agencies, is allowed to fail some fraction of the time."

Case in point, DARPA's investigations in the 1970s into extrasensory perception, or ESP, as a possible means of divining Soviet military secrets. Work into "remote viewing" had already been going on at SRI under a top-secret CIA-funded program when DARPA sent over a group of investigators of its own to check out for themselves whether there was any truth to the claim that a clairvoyant undergoing tests there, one Uri Geller, could manipulate metal objects with his mind or could peer behind the Iron Curtain. They concluded after just half a day's investigation that the claims were nonsense, and dismissed the SRI work as "incredibly sloppy." One of the DARPA researchers, Ray Hyman, a psychology professor at the University of Oregon, chalked up all the attention the clairvoyant was getting to gullibility on the part of the scientists studying him. "They already believed in ESP," scoffed Hyman, who also happened to be an accomplished amateur magician and thus skilled at spotting such sleights of hand as he concluded that the clairvoyant was perpetrating. "They didn't even bother searching him for magnets because they feared that if he felt threatened that way his psychic abilities would be disturbed."

Tony Tether later acknowledged that DARPA's investigations with Geller were only part of a larger effort during the 1970s to get to the bottom of whether remote viewing had any scientific validity. "DARPA spent, for those days, considerable amounts of money because the impact would be tremendous if you could do it." Not to mention the nasty surprise that would result if the Soviets gained useful control of ESP first. "DARPA's original charter," said Tether, "charged the agency to direct and perform 'certain advanced research-and-development projects,' with the primary mission of ensuring the United States would never again be surprised by another nation's technological advancement." But, said Tether, the agency's mission has evolved over time so that now the goal is not just preventing surprise, but also to "create technological surprise for our adversaries." That necessarily means running down some dead ends just to make sure that if they do eventually lead anywhere that no one else gets there first.

A higher-profile area of research with a questionable outcome for DARPA has been the field of ballistic missile defense, or BMD. Following the loss of its space programs, BMD, in the form of a program called Defender, was ARPA's largest budget item through the 1960s. Every possibility, ranging from knocking out enemy missiles in flight with space-borne BBs and ray guns to interceptor missiles designed to hit incoming bombs as they rained down on American cities, has been explored by DARPA (along with many other military offices). In all the decades since these ideas were first explored, no satisfactory solutions have ever been advanced. Effective energy weapons capable of knocking down enemy missiles have yet to be developed, space-based BBs proved too costly to be practical, and interceptor missiles have never demonstrated with any degree of accuracy the ability to separate incoming missiles from the inevitable decoys and chaff.

And even if an effective missile shield could be developed, many outside observers have questioned the advisability of deploying such a thing in the first place. Since one of the most effective deterrents against a nuclear strike is the idea of mutual assured destruction—that is, the notion that any nation whose major cities were attacked with nuclear missiles would immediately retaliate in kind—an effective BMD could actually *encourage* attack. Once a nation found out that an adversary was building a BMD, the argument goes, it would have every incentive to launch a preemptive strike before the BMD became operational, rather than risking being attacked later.

Former ARPA director Charlie Herzfeld scoffs at this argument. Even a less-than-perfect BMD is worth pursuing, he says, because such a system "raises uncertainty on both sides, and uncertainty cannot be overcome by a first strike. So it's very stabilizing. In fact, defenses are highly stabilizing, and people who think they are destabilizing haven't understood the problem. Should I have a little shield just in case you have a sword? The answer is, you bet. Maybe you'll think a lot before you draw the sword. That's the point of defense."

Then, too, numerous advances made in the pursuit of BMD have proven extremely useful in other applications. Take lasers, for example. While try as they might, BMD researchers have been unable to create a "speed-of-light weapon" capable of downing an enemy nuclear missile, they have made great improvements in laser technology in general. Halide excimer lasers operating in the ultraviolet range, developed with a view toward missile defense, have found their way into such varied uses as the manufacture of integrated circuits produced using a process known as deep-ultraviolet lithography, and in LASIK vision-correction surgery.

Pure scientific research has also been helped by BMD efforts. In the early 1960s, ARPA provided funds to build what is still to this day the largest single-aperture telescope in the world. The Arecibo radio telescope in Puerto Rico, which features a 305-meter dish built into a natural crater, has enabled countless discoveries, both within and outside our solar system, including the rotational rate of the planet Mercury, solid evidence for the existence of neutron stars, and the discovery of the first planets outside our solar system. Missile defense researchers hoped to use the telescope to detect the radio signatures of ballistic missiles reentering the atmosphere over North America. That particular line of research proved to be a dead end, but Herzfeld was so taken with the telescope's potential for other, unrelated purposes that he kept it in operation. "Many people wanted me to kill this," he explained later, "but I didn't because I thought there was really very interesting stuff being done—for military and for scientific reasons." Calling the telescope "an amazing success," Herzfeld included it among the ARPA projects that "happened serendipitously that were not of top importance but had wonderful side effects."

Those wonderful side effects that DARPA was so good at creating led to a shift in focus for the agency after the fall of the Soviet Union in 1991. A name change in 1993—dropping the *D* to again become ARPA—reflected a new focus on deliberately fostering civilian as well as military uses for the technologies it developed. The name change was brief, however. In 1996, ARPA again became DARPA. "What had happened in the nineties was that it was unclear what the mission was," Tether told me. And it wasn't just ARPA/DARPA that was flailing. With the collapse of America's cold war enemy, "It was unclear to the whole Department of Defense what *their* mission was," said Tether.

"DARPA was used as an experiment in the early nineties," he explained, "to go off and do this dual-use technology. You know, technology that was good for the commercial industry as well as the military." As far as Tether was concerned, making what had always been an incidental and serendipitous by-product of DARPA's work, commercial technologies, into DARPA's explicit mission made no sense at all. "You get kind of schizophrenic," he said, describing the mind-set of DARPA's managers during that time. Getting the agency back on track as a purely military operation was one of the missions Tether was given when he came in as director of DARPA in 2001.

He became director through a circuitous route. After earning his Ph.D. in electrical engineering from Stanford University in 1969, he founded a defense contracting company in Palo Alto, California, called Systems Control, which did a lot of number crunching to determine the best possible ways to allocate resources within both military and commercial organizations. The company's mission, Tether told me later, was "developing technologies that quite frankly people didn't know they needed, so we were constantly having to convince people that what we had was something that they wanted, even though they didn't know they wanted it." Excellent training, Tether realized later, for a future head of

DARPA. In fact, Tether put that experience of convincing others of the utility of new technologies right on a level with actually knowing how to create that technology as one of the most important qualities for a DARPA director. For that reason, he named an even earlier job he had as an important part of his training for his future work. "I was a Fuller Brush man," he told me. "A Fuller Brush man has that same job: You go knock on the door. You've got a few seconds. You have a good product, but now you have to convince this person that they want to buy a brush when they really may not know that they need to buy a brush."

After British Petroleum bought Tether's technology company in the late 1970s, Tether became the head of the National Intelligence Office in the office of the secretary of defense, and discovered that he liked government work. So when President Reagan came into office in 1981, Tether fought to stay on in his current job, but, as he told me, "even though I tried to convince everybody that I had voted for Reagan more times than anybody else in Washington because I came from California, it didn't matter, because the job I was in was one of those jobs that changed with administrations." Just like the job of DARPA director, which is why although Tether's tenure was to become the agency's longest, he figured it would last only as long as the presidency of George Bush the younger.

A reprieve from Tether's impending dismissal from the National Intelligence Office came in the form of an invitation from the then DARPA director, Bob Cooper, who wanted Tether to head up DARPA's Strategic Technology Office, or STO. The job seemed a natural for Tether, who had worked with DARPA as a contractor through Systems Control. If anything, he enjoyed his first stint at DARPA even more than his previous government job. To participate in the creation of breakthrough technologies, even to be directly responsible for them, was a rush like no other. Citing just one example, he described to me the meeting that led to the creation of GPS receivers small enough to slip in a pocket—devices that today are quickly becoming as commonplace as the cell phones of which they are increasingly an integral part.

It was then the early 1980s, and the modern GPS satellites were still being launched. Receivers for picking up signals sent by those satellites 22,300 miles straight up were still heavy, bulky affairs. The 17-pound backpack-mounted units called Manpacks were so big that only one soldier in a given outfit could carry one, since that person would have to forgo a significant amount of other gear. Often the thing was mounted on a truck or a helicopter, reducing its utility even further. In either case, the size and weight of the GPS receiver made it impossible for more than one or two soldiers in a unit at a time to know with any precision where exactly they were.

STO program manager Sherman Karp thought he could improve on that situation, and he walked over to Tether's office. "Look," Karp said, as Tether later recalled, "because of the microelectronics that we have developed, I believe that we can take that big receiver and shrink it down to a very small size."

Tether asked the obvious question: "How small?"

Karp looked around Tether's office. Tether was a smoker at the time, and Karp's gaze settled on the pack of cigarettes on Tether's desk. "Well," Karp ventured, "I can bring it down to the size of your cigarette pack."

"Okay," Tether said immediately, "that's what you have to do."

Perhaps at that moment Karp realized how his impulsive response might paint him into a corner, and he tried to backpedal a bit. "Do I have to have the battery there, too?" he asked.

Why not, thought Tether, go for broke. "Yes, it has to contain the battery," he replied.

When Iowa-based Rockwell International, the makers of the Manpack, got the assignment from Karp to build the cigarette pack–size GPS, they had a hard time taking it seriously. "What the customer wanted, and we joked about it," recalled Rockwell engineering manager Ron Coffin later, "was a handheld receiver the size of a Virginia Slims cigarette pack." DARPA might as well have been asking for a long-range bomber the size of a Buick. "No kidding," said Coffin. "That's how the customer described the requirement." Coffin had one of his engineers make up a drawing of a Virginia Slims pack–size receiver, with an antenna sticking out of it in place of the cigarette that the brand's ads typically showed jutting out of the pack. The receiver's screen displayed the Virginia Slims slogan: "You've come a long way, baby." Coffin had the picture delivered to his lead technical engineer as a depiction of what they were supposed to build.

Back at DARPA headquarters, neither Karp nor Tether actually had any idea whether it was indeed possible to shrink a GPS receiver down to the size of a cigarette pack, but this was just the kind of high-risk, high-payoff investment in technology that DARPA had been set up to make. Said Tether later, "The projects that we do are typically those projects [where] the idea could end up with a great capability, but there's extraordinary little data to prove that the idea could be built. . . . DARPA's role and uniqueness is that we will take a bet on that idea where other people will not."

And as it turned out, that particular idea did indeed pay off, and like stealth aircraft, said Tether, it became a game changer. "It led us to precision missiles, because we had GPS as the way to navigate. So, on the military side, it was fantastic. On the commercial side, it was just as fantastic. I mean, we now can . . . have our cars drive themselves. We now have tractors that fertilize fields, all running on a GPS. All from a person who had an idea."

It's no wonder that Tether described his departure from DARPA in 1986 as "a great decompression." He explained, "When you're in Washington, you tend to think you are the center of the world—until you leave the Beltway and find out that nobody else cares."

He went to Ford Motor Company as chief technology officer, before leaving in 1992 for the Advanced Technology sector at Science Applications International Corporation, a major defense contractor, and from there he went on to found a technology consultancy firm in California called the Sequoia Group, with clients in both government and the private sector. Not incidentally, one of his clients was DARPA, so he still had his finger on the pulse of the agency. It was in this latter capacity that he was tapped by the administration of President George W. Bush in 2001.

By then he'd grown fond of his burgeoning paycheck. As Tether put it to me later, "I tried really hard not to get a job," because he did not relish the idea of taking a "financial bath."

Still, at the insistence of people he knew in the new administration, particularly an official named Edward "Pete" Aldridge, who was to be undersecretary of defense for Acquisition, Technology and Logistics, Tether sent in his résumé, and there followed a series of interviews at the Pentagon. "They were pretty high-level jobs," Tether told me later. "They were all jobs that you had to go be confirmed by Congress for. And I just avoided them. I thought, 'My God, you know, I don't really want to do that.' But I didn't want to offend anybody, so I just did my best to convince people I wasn't qualified for it."

Around this time, Tether met Gordon England, who was in line to be the secretary of the navy. "And I said, 'Gordon, you know I'm really not qualified for this job you want me for, blah, blah, blah.' And that irritated him." England wouldn't be brushed off so easily. "He finally said, 'Well, look, dammit, I know that you are definitely qualified to be the director of DARPA.'"

That did it. Tether had to admit that, yes, he thought he was qualified for that job. "I got trapped," he told me later, "because they were right: I was qualified to be the director of DARPA," and try as he might, "I couldn't think of why I wasn't." As reluctant as he initially was, Tether said, "I don't regret a moment of it, ever since I decided I could do it." Technically a presidential appointee, Tether was actually hired by Aldridge. "He said, 'Okay, you're the director of DARPA,'" Tether told me with a chuckle.

A final hurdle to jump before settling into his new office was a job interview with the new secretary of defense, Donald Rumsfeld. It was a more or less pro forma interview. In fact, for a while during the interview, Tether wasn't even sure Rumsfeld knew why Tether was there. "It was supposed to be for fifteen minutes, and I got there and it lasted for, God, I don't know, about forty-five minutes." During most of that time, Rumsfeld spoke generally about the problems he wanted to solve and about the changes he wanted to make at the Department of Defense. Finally, as the interview was drawing to a close, Rumsfeld got to the point. "You know," he told Tether, "I'm here trying to transform the government, the DOD, into something new. But I don't want to transform DARPA into something new. I want you to make it like it was."

"There was a concern that DARPA had kind of lost its way," Tether explained to me later. "There was a concern that DARPA was not the freewheeling kind of DARPA that Pete [Aldridge] and I both knew in the eighties." The kind of DARPA that Rumsfeld had also known when he was secretary of defense in the Ford administration in the 1970s. Back in those good old days, Tether explained, "the program managers always had the DARPA director on the verge of being fired for something." Back then, people ran around the office "with their hair on fire . . . thinking crazy-ass things." Meaning that they were thinking so hard and for so long that their brains heated up enough to set their hair alight. That was the DARPA Rumsfeld wanted Tether to bring back. He wanted Tether to return the agency to its roots as a free-spirited home for mavericks and mad dreamers operating on the barely respectable fringes of Pentagon culture, the only place where truly creative thought could thrive. Tether couldn't have agreed with this direction more, and now that he had re-signed himself to returning to a government pay scale, he was eager to take on the challenge.

"Oh, by the way," Rumsfeld said as Tether got up to leave. "Space is important." And he handed Tether a copy of the Report of the Commission to Assess United States National Security Space Management and Organization, also known simply as the Rumsfeld Commission report because Rumsfeld chaired the commission until his appointment as secretary of defense. Charged with determining how best to organize America's assets in space, the commission also assessed the *importance* of those assets and their vulnerability to attack.

In no uncertain language, the report, released in January 2001, warned that the United States was highly vulnerable to a surprise attack in which the GPS, communications, and spy satellites on which the U.S.

military was utterly dependent for navigation, communications, and intelligence-gathering could be wiped out at a stroke. "The U.S.," the commission reported flatly, "is an attractive candidate for a 'Space Pearl Harbor.'"

"We know from history," said the report, echoing the ABMA's General Medaris of a generation before, "that every medium—air, land and sea—has seen conflict. Reality indicates that space will be no different." Recognizing the "sensitivity that surrounds the notion of weapons in space for offensive or defensive purposes," the Rumsfeld Commission nevertheless asserted that to ignore the space peril "would be a disservice to the nation," and the U.S. government should take steps to "ensure that the President will have the option to deploy weapons in space to deter threats to, and if necessary, defend against attacks on U.S. interests."

The report stopped just short of recommending the creation of a space force. The commission did, however, recommend creating an undersecretary of defense for space, intelligence and information, and consolidating space warfare activities under the command of the air force as a dedicated space corps. Most relevant to Tether as incoming DARPA director, the commission placed special emphasis on keeping America ahead of the rest of the world in space technology. "In particular," said the report, "the government needs to significantly increase its investment in breakthrough technologies to fuel innovative, revolutionary capabilities." The commission report even mentioned DARPA by name in its recommendations, saying, "The Secretary of Defense should direct the Defense Advanced Research Projects Agency and the Services' laboratories to undertake development and demonstration of innovative space technologies and systems for dedicated military missions."

Now Rumsfeld was telling Tether that he wanted DARPA to pick up where it had left off when it had ceded its space mission to NASA more than forty years before. His course set, Tether got down to work. Like his predecessors, he felt he had no time to lose; he anticipated staying at DARPA for only a couple of years, just long enough to steer the agency back in the right direction. Most previous directors had served terms of no more than three years, and Tether figured he'd be no different. What he couldn't have reckoned on was 9/11.

Tony Tether was in his office at DARPA's Arlington, Virginia, headquarters on the morning of September 11, 2001, when five knife-wielding terrorists flew American Airlines Flight 77—a Boeing 757-223 loaded with passengers and enough fuel for its planned cross-country flight—directly into the western face of the Pentagon. The aircraft exploded into a brilliant orange fireball 200 feet high, killing all 64 people on board and 125 military and civilian personnel in the building. Tether already had the television in his office on to watch news of two other hijacked airliners that had slammed into the twin towers of the World Trade Center in New York less than an hour before, and soon the screen filled with images of the devastation much closer to home. He didn't need to watch the television, though. He needed only look out his own office window to see the black smoke pouring into the sky just four miles away.

For DARPA and the Pentagon, as well as for the nation, that terrible morning changed everything. For Tether, only a few months on the job, it was his trial by fire. As he told me later, all thought of leaving the agency anytime soon evaporated. "It wasn't a conscious decision," he said. "I didn't wake up on September twelfth and say, 'Okay, I'm here forever.'" But he and others of his colleagues in government felt an overwhelming desire to serve, to *do something* to help the country defeat this new, amorphous, largely faceless, nationless enemy—an enemy, it seemed, that would require novel weapons and tools to defeat.

"I can remember standing in my office and looking at the TV, watching the towers collapse, and looking out the window and seeing the black smoke coming from the Pentagon," Tether told me. "And I knew that at the end of the day that we were going to find that we had all the data we needed to know about this." Tether was certain that some government office somewhere had at some time intercepted a telephone call, e-mail, or financial transaction that, if properly interpreted and brought to the attention of the right people, would have allowed authorities to stop the attacks of September 11 before they happened. What was lacking was neither the people nor the leadership at those agencies, but the data-mining tools needed to connect the dots before 9/11. "So we started a major thrust to never let that happen again."

This "major thrust" was more than just a single program. It was too big for that. Instead, Tether started an entire program office called the Information Awareness Office in 2002. Its most prominent program was Total Information Awareness, or TIA. The idea, as Tether later put it, was to "break down all the stovepipes," to give government officials the tools they needed to sift through the enormous amount of data already flowing into their offices and detect patterns and flag suspicious activities for human analysts to act upon. As it turned out, Tether's first major undertaking as DARPA director almost became his last. "The agency got into a lot of controversy," as he explained it. "I mean an enormous amount of controversy."

Tether's first controversial move was bringing in Admiral John Poindexter, a physicist by training, as director of the Information Awareness Office. "And to this day," Tether insisted to me in 2008, "John Poindexter was the absolute right person to run the office. You know, I still believe that." But Poindexter brought with him his own brand of controversy, and just by his presence, drew scrutiny to Tether's new office. Poindexter had been President Reagan's national security advisor in the mid-1980s when the story broke that high-level U.S. government officials had been selling arms to Iran to obtain the release of American hostages held there and to illegally fund antigovernment forces in Nicaragua known as Contras. Poindexter was implicated in what became known as the Iran-Contra scandal, and in 1990, he was convicted of conspiracy, lying to Congress, obstruction of justice, and destroying evidence. And then, the following year, his convictions were overturned on appeal after his lawyers successfully argued that he had been granted immunity in previous testimony to Congress, and that the testimony had been improperly used against him. That was good enough for Tether, who concluded, "He never really was convicted of anything." But Tether did have to admit to me that Poindexter came with a certain amount of baggage that did not serve DARPA well.

The beginning of the end of TIA and Poindexter as a DARPA program office director was a scathing article by William Safire titled "You Are a Suspect," which appeared in *The New York Times* on November 14, 2002. Under TIA, Safire told his readers, "Every purchase you make with a credit card, every magazine subscription you buy and medical prescription you fill, every Web site you visit and e-mail you send or receive, every academic grade you receive, every bank deposit you make, every trip you book and every event you attend—all these transactions and communications will go into what the Defense Department describes as 'a virtual, centralized grand database.'"

Tether was actually able to laugh about the article as he talked about it years later. The idea that DARPA wanted to look into every U.S. citizen's private records was nonsense, he told me. What he and Poindexter wanted to do, he said, was to develop sophisticated methods for sifting through the terabytes of data already being amassed by the various law enforcement, military, and intelligence organizations to spot suspicious patterns that might correlate to terrorist activity—before that activity turned deadly. "We never really meant that we are going to sift through the cyberspace of the United States. We were really always looking at the normal amount of information that the intelligence agencies collected as matter of fact. But, boy, you know, I never could really convince people of that."

One of the unconvinced was Electronic Privacy Information Center director Marc Rotenberg, who said in a *Washington Times* article that appeared the day after Safire's article that regardless of DARPA and Poindexter's actual intentions, TIA represented a slippery slope toward the loss of important freedoms by the American people. "They think the technology is about catching terrorists and bad guys, but these systems can capture a lot of data at different levels without oversight, judicial review, public reporting or congressional investigations. I can't think of a good countermeasure that would be good to safeguard civil liberties in the United States."

As the controversy grew increasingly heated, Congress began an investigation into the privacy concerns raised by critics of the Information Awareness Office, suspending the office's activities in the meantime. With the IAO in a fight for its life, DARPA issued a report in its defense, at the same time changing the name of the controversial program from Total Information Awareness to Terrorism Information Awareness, in an attempt to quell fears that DARPA wanted to engage in domestic spying. "This name created in some minds the impression that TIA was a system to be used for developing dossiers on U.S. citizens," said the report. "That is not DOD's intent in pursuing this program. Rather DOD's purpose in pursing these efforts is to protect U.S. citizens by detecting and defeating foreign terrorist threats before an attack." But it was too late. The damage to public perception had been done. The Defense Appropriations Act of 2004, signed into law by President Bush in October 2003, canceled TIA and disbanded the IAO.

During the height of the controversy, Tether received a compliment from Rumsfeld, as Tether put it to me later, "only the way Rumsfeld could give you a compliment." The secretary of defense sent a high-level messenger over to Tether's office to tell him, "The boss wants you to know that you are getting close." As in close to getting fired. Tether laughed as he relayed the story to me. It was a tacit acknowledgement from Rumsfeld that Tether was indeed following the mission he had been given—to shake things up at the agency and get it back to the freethinking of the good old days.

Tether and the agency managed to weather the storm, even if Poindexter had to be sacrificed, but it was touch and go there for a while. "I was really worried," Tether admitted to me. "I was really worried about what could come out of this, so I stuck it out." At stake was the independence of an agency that thrived with relatively light oversight. During its investigation into TIA, Congress had considered placing stricter controls on DARPA, to ensure that *all* of its programs stayed within appropriate bounds. In the end, DARPA wriggled clear with only the loss of TIA and IAO, and eventually the agency went back to the relative obscurity in which it best operated. "The controversy went away, and we kind of got over that," Tether told me. "By the time we got over that, five years had passed." Already he'd served longer than any previous DARPA director. "And so as long as I was here for five years, I decided, 'Well, I might as well stay for the fiftieth [anniversary].'"

When I met with Tether in his office in 2008, he figured he was definitively at the end of his term; with the changing of the presidential administration in 2009, he would have to go, no matter which party came to power. "You know," he reflected, "I used to always say that the greatest thing about DARPA is that no one's been there long enough to screw it up. Unfortunately, I've been here so long there are people who have said to me, 'Hey, you remember when you said that, Tether? Well, you know, aren't you getting close to that time?'"

Not everyone was delighted with the way Tether ran the agency. Some researchers felt that Tether had pulled DARPA too far in the direction of producing practical applications in the near term. "He's disemboweled DARPA's long-term scientific effort," one researcher complained. He aired his concerns to me on condition of anonymity. "Don't get me fired," was the way he put it. He, along with some of his colleagues at other institutions, felt that Tether had turned DARPA into a "technology development center." Which meant to him that DARPA under Tether had abandoned the long-term research that had been within its purview previously to focus on short-term assignments that could produce immediate results. This, in turn, had played havoc with the researchers, particularly at universities, which now had to constantly scramble after money and which could no longer promise graduate students continuous work. "It's on again, off again," said my source, so that project leaders "can't even tell graduate students that they'll have a project to work on" from one year to the next. He couldn't have been stronger in his criticism, calling DARPA's direction under Tether a "catastrophe."

My source wasn't the only one who disagreed with Tether's direction of DARPA. No less a figure than Leonard Kleinrock, the head of the ARPANET project at UCLA, actually stopped taking DARPA funding in the 2000s because the money would have come with the stipulation that he not use graduate students in his work. The restrictions went along with a tightening of money awarded to universities as DARPA shifted to shorter-term projects with more immediate (and thus more sensitive) military value and away from the kind of open-ended pure research like the kind pursued by Engelbart at SRI in the 1960s and '70s. Without funding—much of which happened to come from DARPA—for that kind of research, many of the technologies we take for granted today, such as the computer mouse and the Internet, might have died on the vine. "Virtually every aspect of information technology upon which we rely today bears the stamp of federally sponsored university research," Ed Lazowska, a researcher at the University of Washington, told the *New York Times* in 2005. "The federal government is walking away from this role, killing the goose that laid the golden egg."

Tether himself remains unapologetic about DARPA's focus on applied research. "Even the ARPANET, or Internet as it is known today, had an application in mind and that application directly and indirectly drove the research," he told me in early 2009. He defends the direction in which he pushed DARPA during his tenure. "Open-ended research is categorized as 6.1 research in DOD and is defined as research without an application in mind." DARPA's role during his directorship, he said, was to mine this research for ideas that could be applied. As he put it, "DARPA program managers constantly search open-ended research for new inventions, and fund them if organizations such as NSF—whose primary purpose is open-ended research—do not." But even with that focus, "DARPA's 6.1 budget roughly doubled over the past eight years, whereas DARPA's total budget only increased 50 percent. In financial terms, DARPA's open-ended research outperformed the total DARPA program. Statements that open-ended research decreased over the past eight years are at odds with these facts."

That many expected DARPA to play a major role in fostering innovation for its own sake was a testament to the agency's success. It certainly wasn't part of the agency's charter, however, and Tether was under no obligation to serve any but military interests. "There are pressures and demands on DARPA to be relevant," was the way Internet pioneer and former DARPA program manager Bob Kahn put it to the *Times*. "People think it should stay the same, but times have changed."

Still, DARPA couldn't help but stay relevant, as long as it continued to push the envelope in such diverse fields as medicine and transportation. By his own assessment, Tether had succeeded in his mission to keep the entrepreneurial spirit alive at DARPA, and if he ruffled a few feathers along the way, so be it. Skirting the edge of danger, he figured, was one of the marks of a successful DARPA director, and he expected his program managers to do the same. Program manager Kirkpatrick, for one, embraced the challenge. "We have a sort of mutual defense treaty, I guess is the best way to put it," Kirkpatrick told me of his relationship with Tether. An "'if I get fired, he gets fired' kind of deal."

In April of 2008, I joined some sixteen hundred DARPA program managers, directors, and contractors past and present for a low-key black-tie fiftieth-anniversary celebration for the agency at the Washington Hilton. "Over the years," Tether said, addressing the crowd, "DARPA has responded to issues of national importance with new ideas and technology that have changed the way wars are fought." And, he added, "changed the way we live." As if those gathered needed to be told.

Or maybe they did. With much of what they did cloaked in secrecy, and even information about their nonclassified projects still tightly controlled, they toiled in obscurity. It must have been nice to get a little acknowledgement of the importance of their work, even at a private function.

Realizing that he'd not likely have another chance to directly address so many DARPA luminaries in one place, Tether also took the opportunity to bid them farewell as director. He did it in the most understated way possible, with an unadorned "Good-bye."

Vice President of the United States Dick Cheney arrived after the banquet portion of the evening and told the gathering, "Anyone who seeks the very definition of high intellectual standards, creative energy, and hard, persistent effort, will find it at DARPA. It's a huge credit to the Defense Department and to the nation."

After dinner, the group mingled over cocktails. Everyone stayed sober, and after a discreet interval, quietly slipped, one by one, into the springtime night, causing hardly a ripple or a second glance amid the tourists, business people, and other government workers among the guests of the hotel.

Much to Tether's surprise, DARPA's fiftieth anniversary year came and went, and he found himself still in the director's chair. When Barack Obama succeeded George W. Bush as president in 2009, he made the unusual decision to keep his predecessor's secretary of defense, Robert Gates, in place. Tether was one of those under Secretary Gates who kept his job in the new administration—for a while. "Dr. Tether will be here after January 20," Jan Walker e-mailed me in response to my query about what Tether's status would be after Inauguration Day, "and there's no formal date on which he plans to leave."

However, this reprieve turned out to be short lived. Tether had to leave DARPA on February 20, and mechanical engineer Regina Dugan, who had been a DARPA program manager in the late 1990s, took office as DARPA's first woman director on July 20.

"DARPA is envied because of the capabilities it has brought to realization far in advance of anyone else," Tether told me after his departure. "This is due to its focus on bridging the gap between open-ended research—the far side—and the near-side research, which improves what exists today. DARPA program managers hired during my tenure all had the capability when they joined to evaluate open-ended research and determine what new capability was now possible. By the time they left, they learned techniques on how to articulate to potential users why they needed this new capability even if it meant the user giving up current programs and doctrine. This combined ability allowed them to develop and transition technologies that provided new capability for the U.S. and saved many lives. I couldn't be prouder of them and am honored to have been at DARPA when they were there."

The ability to continually renew itself is, of course, one of the agency's strengths. But it is also true that many areas of DARPA research span more than one directorship. DARPA-funded researchers who are focused on the conquest of space, for example, can only hope that their work will be among the projects started by Tether to continue under subsequent directors.

THE FINAL FRONTIER

THE ROAR SHATTERED the stillness. This late summer day in Hampton, Virginia, in 2008 was hot and bright, and I had to shield my eyes and squint as I looked up at the pair of F-22 Raptors climbing out of Langley Air Force Base. The delta shapes of the stealth fighters looked like toys against the blue sky, with sandy gray undersides and twin out-thrust vertical stabilizers. One of them peeled right, and the roar got louder, making speech with the NASA public relations rep standing beside me in the parking lot of NASA's Langley Research Center impossible. The raptors were the final word in fighter technology, incorporating the state of the art in stealth tech developed by DARPA and its contractors, able to penetrate enemy radar at twice the speed of sound to deliver their lethal payloads in precision-guided strikes. Yes, they were fast all right. But not nearly as fast as the experimental machine I was here to see.

The Langley Research Center had been built in 1917 here on a spit of land near Newport News on the shore of the Chesapeake Bay. It began life as the Langley Aeronautical Laboratory at the dawn of the age of aviation, and it was the nation's first laboratory for civilian aeronautical research. As such, it was also NASA's first research center, given its current name when the new space agency took over. The center had been massively upgraded at the start of the space race in the late 1950s and early 1960s, and seemed to have changed little since. There, suburban-style streets were lined with low-slung buildings whose backyards were filled with gigantic pipes and tanks feeding the research center's wind tunnels. After three other carefully selected journalists joined us, we all headed to the Eight-Foot High-Temperature Tunnel, the largest one in the complex capable of reproducing the heat and howling wind conditions of hypersonic flight—that is, past five times the speed of sound.

A group of aerospace managers, engineers, and technicians met us in the building that housed the tunnel to give us a briefing on a little-known project that could revolutionize air and space travel: the X-51A. The goal of the project was to build an airplane capable of sustained flight at hypersonic speeds—a realm that until very recently was the exclusive domain of rocket ships. Air-breathing engines capable of that kind of speed, if proven practical, could precipitate as big a change in aviation as the advent of the jet engine itself.

An ordinary jet engine depends on fast-spinning turbines to compress incoming air to the pressures needed to ignite jet fuel at a high enough temperature to allow the expelled exhaust to thrust an aircraft forward at high speed. Generally, the faster the turbine spins, the greater the resulting thrust, and thus the faster the aircraft can go. But turbines constructed of the highest-strength alloys available can spin only so fast before heat and friction tear them apart. That's why the fastest jet plane ever built, the SR-71 Blackbird, could hit a top speed of only Mach 3.3. An experimental engine design called a scramjet (short for supersonic combustion ramjet), however, had a theoretical top speed somewhere in the neighborhood of Mach 15. It was this design that the X-51A project sought to perfect.

Scramjets differ from ordinary jets in one very important way: their guts contain no moving parts. Instead of relying on turbines to compress incoming air for ignition, scramjets rely on incoming air to be precompressed by the sheer speed at which the aircraft is already moving. Blowing through the scramjet engine, the air gets squeezed and superheated by nothing more than the geometry of the engine before hitting the precisely injected fuel, which then explodes with enough propulsive power to send the aircraft faster than any jet plane has gone before. The exact geometry of the engine and the precise manner and timing of the fuel injection required for the alchemy of sustained scramjet flight were only now just beginning to be understood. The difficulty of the challenge has been likened to lighting a match in a hurricane, but the X-51A engineers thought they just about had the problem licked. If so, their X-51A engine would for the first time give an experimental aircraft the ability to maintain hypersonic flight.

Ultimately, scramjet-powered airplanes could fly passengers halfway around the world in four hours. Compared to the twenty-five-hour flights now required to travel that distance, scramjet technology has the potential to revolutionize any activity requiring long-distance travel. But there are a lot of intermediary steps to getting there. The first long-range hypersonic research aircraft are likely to do little more than fly far and fast, and then simply crash. That pretty much describes the flight profile of a missile, and that's exactly the first application the Air Force Research Laboratory, or AFRL, and its partner in funding the X-51A, DARPA, sees for scramjets. Instead of launching a massive rocket to hit a target a continent away, the air force could instead drop a scramjet-powered cruise missile off a fighter plane to hit any of those targets with very little notice and much greater effectiveness.

While there's no theoretical speed limit for rockets, they are hobbled by the need to carry their own oxygen supply. That's a serious handicap. Consider that while the space shuttle has to carry a burdensome 1.3 times its weight in liquid hydrogen fuel, it also has to carry a downright staggering 8.5 times its weight in liquid oxygen to get that fuel to burn on its way to its Mach 25 orbital speed. That much extra baggage means more tankage, and hence even more weight to carry, resulting in a vicious cycle of increased fuel and oxidizer needed to carry extra weight, and thus more tankage. The big orange tank the shuttle rides off the launch pad is filled mostly with liquid oxygen.

Relieved of the burden of all the oxidizer a rocket has to carry in order to burn its fuel, an air-breathing engine has only to propel the weight of the vehicle and its payload, plus jet fuel. Thus it takes a much smaller vehicle to carry the same payload. But the theoretical advantages of a hypersonic jet over a rocket don't stop there. Hypersonic jets are also theoretically much more fuel efficient than rockets.

"With an air-breathing system, you get a lot higher specific impulse, which is the measure of thrust versus your mass flow," DARPA hypersonics program manager Steven Walker explained to me. "A typical ISP, or specific impulse, for a rocket is around two hundred seconds, and with an air-breathing ramjet, scramjet system, you can get around one thousand." That much greater efficiency means that a scramjet can carry less fuel than a rocket as well as no oxidizer at all, allowing yet more weight to be devoted to actual payload—or that the payload could travel farther on a given volume of fuel. All of which results in faster and higher-flying missiles in smaller packages. It's no wonder DARPA wanted hypersonic cruise missiles. But they were a long time coming.

Scramjet fever broke into the mainstream back in 1986, when President Reagan chose his State of the Union address as the venue to announce the formation of a concerted national effort to build a working scramjet. This new vehicle was to be called the National Aerospace Plane, or NASP. Delivering his speech on February 4, just a week after the Space Shuttle *Challenger* became a fireball after liftoff and took the lives of seven astronauts, Reagan sought to reaffirm his nation's commitment to a continued presence in space. "We are going forward with our shuttle flights," he said. "We're going forward to build our space station. And we are going forward with research on a new Orient Express that could, by the end of the next decade, take off from Dulles Airport, accelerate up to twenty-five times the speed of sound, attaining low Earth orbit or flying to Tokyo within two hours."

It was during this heady time for hypersonics research that DARPA's Steven Walker began his career as an aerospace engineer fresh out of college—Notre Dame, in Indiana—at Wright-Patterson Air Force Base. Since he had grown up in Dayton, Ohio, the job gave him a welcome chance to come home again. But his work at Wright-Patterson was strictly about blazing new territory. Walker went right to work on the NASP project, plunging into what looked like the wave of the future, hypersonic air-breathing engines.

The NASP, also known as the X-30, was based on research conducted from 1982 to 1985 for a classified DARPA program called Copper Canyon. Unfortunately, it never got off the ground. Part of the problem was that it was to be all things for all comers. It was to be the vehicle that succeeded the space shuttle in carrying satellites and astronauts to orbit. It was to be the pinnacle of air force bomber technology, and it was to be a hypersonic passenger transport for point-to-point travel on Earth. Ultimately it fulfilled none of those expectations; the technical challenges simply proved to be too enormous. Not the least of the problems was that Mach 25 goal. The actual maximum speed of a scramjet, researchers discovered, was probably more like Mach 15. Certainly creation of an airplane capable of that speed would be a great leap forward for aeronautics, but in light of the high expectations placed on the NASP, it looked positively poky. Congress and President Clinton quietly canceled the project in 1994.

After NASP lost its funding, Walker had no choice but to switch his focus to more conventional projects, which included the F-22 Raptor. He also attended the University of Dayton part time to earn his master's degree, which eventually led him back to Notre Dame for his Ph.D., on the air force's dime. Meanwhile, still getting a modicum of government funding, scramjet researchers went underground. And out of the glare of the public spotlight and freed from unreasonable expectations, they began to make real strides. "We didn't stop our scramjet research," Charlie Brink, scramjet program manager at the Propulsion Directorate of the Air Force Research Laboratory, or AFRL, told me. "We reevaluated it and said, 'You know, now that we're not trying to make a Mach-zero-to-twenty-six vehicle take off from a runway, let's take the technical problem and break it down into more manageable chunks.'"

In 1997, the year he earned his Ph.D., Walker jumped at the chance to take over the job of overseeing aeronautical and hypersonics research at the Air Force Office of Scientific Research, or AFOSR, in Arlington,

Virginia, when his predecessor moved to DARPA. It was while working at AFOSR that Walker took what turned out to be a brief assignment at the Pentagon, working for the director of the Department of Defense Research and Engineering, or DDR&E. Walker had been working for the DDR&E for all of a month and a half when terrorists flew American Airlines Flight 77 into the Pentagon's western face amid an explosion of jet fuel. He was on the other side of the mammoth building, the world's largest, and so he neither saw nor heard the explosion that was to alter the course of his life.

After the initial shock began to dull and Walker and his colleagues at the Pentagon went back to work, the first thing they tried to figure out was how this event would change the way they conducted the business of war. "One of the issues that popped up to everybody after 9/11 was: we need faster missile platforms than we currently had," Walker told me. Platforms that could launch missiles fast enough to strike anywhere in the world on short notice. "For me, and the sense in the building was at that time, the cruise missiles we had were very capable, but they were all subsonic." It was going to take weeks to set up U.S. military bases in Afghanistan, home of Osama bin Laden, who had masterminded the 9/11 attacks. "So the thinking was," Walker told me, "if we had other options that could go farther, faster, we could have done more in the early days of that conflict."

"So hypersonics sort of came back to the fore in my thinking," said Walker. Hypersonics, which he had all but given up for dead back in the days of NASP, became his renewed passion. The DDR&E supported Walker in his return to air-breathing hypersonics, and put him to work trying to peer into the future. What could be accomplished in this area in the next five years? Ten years? Fifteen? Walker called the plan he developed the National Aerospace Initiative, or NAI. The NAI drew on the expertise of the army, air force, and navy, as well as DARPA, to lay out a roadmap for hypersonics development. "Basically the plan was start with expendable missile-like demonstrators—demonstrate materials and propulsion technologies that would enable a ten-minute, cruise-expendable missile application—and move into reusable." *Reusable*. The magic word that means a craft that will go far beyond the realm of mere missile. The first step in that direction would be hypersonic aircraft. "And then the last phase," Walker told me, "would be to develop hypersonic platforms for access to space." It was the dream of the NASP—and kept alive by underground scramjet researchers everywhere—that Walker dared incorporate as part of the NAI. The idea was to use a scramjet as the first stage of a rocket that would take its payload out of the atmosphere. Even with the Mach 15 speed limit, such a system could still get a satellite or even a crewed spaceship to space more affordably and with less logistical hassle than a straight-up rocket. But Walker would get there one step at a time, instead of going for broke in one program, as NASP had tried to do.

Walker and his colleagues had learned a valuable lesson from the failure of NASP, which Walker continued to believe was a good program and had been on the right track. "We developed a lot of advanced technologies that we're using today, actually," he told me. "The one major mistake [NASP] made is: we're going to do all this tech development and then integrate it all in one vehicle and fly the final vehicle." In other words, NASP bit off more than it could chew. It had been sold as a program not only to develop all the technologies needed to build a working reusable air-breathing hypersonic vehicle, but then actually to fly that vehicle as well. That simply wasn't going to happen on the budget and within the time frame allotted to it. It had been bound to fail.

Breaking the problem down into more manageable chunks resulted in separate programs for building small-scale engine prototypes, exploring the aerodynamics of hypersonic flight, and more. Small unmanned vehicles were the word of the day. This approach started to pay off in the 2000s. David Van Wie, a longtime scramjet research scientist at the Applied Physics Laboratory at Johns Hopkins University, summed the situation up for me. "What you're seeing now is a transition of the technology out of the laboratories into the flight-test domain," he said. "It's really to the point that people who work in the field feel that they're ready to take the steps into flight test, experimentation, and demonstration."

First out of the lab was the HyShot program of the University of Queensland, Australia. In 2002, researchers at the Centre for Hypersonics achieved the first powered flight of a scramjet engine by shooting a small scramjet engine out of the atmosphere on the tip of a sounding rocket. After jettisoning the rocket, the scramjet fell nose first back to Earth, building up enough speed in the process to allow the engine to light up at an altitude of 20 miles. Five seconds and 7.6 times the speed of sound later, the engine slammed into the Australian Outback, having transmitted crucial test data to researchers back on the ground. After NASA's unmanned X-43A did HyShot one better in 2004, by launching horizontally from an airplane and reaching Mach 9.6 during its 10-second flight, HyShot's successor at the University of Queensland, HyCAUSE, cracked Mach 10 for 3 seconds to set a new record for non-rocket-powered flight.

Meanwhile, Walker's work for the DDR&E was right up DARPA director Tony Tether's alley. And the more Tether got to know Walker, as Walker did his research and made the necessary connections within the Department of Defense to put it together, the more he liked what he saw: a dedicated researcher, quiet about his accomplishments, yet absolutely driven and fearless in his pursuit of knowledge of the unknown. In 2002, Tether invited Walker to join the agency. Specifically, Walker told me, "He asked me to come over and start a reusable hypersonic program for him."

The chance not only to research the possibilities of hypersonic jets, but also to actually manage efforts to build real working hardware in pursuit of his goals was too much for Walker to turn down. "At AFOSR, I sponsored basic research in high-speed boundary layer transition, some scramjet propulsion work," Walker said, "but it was all at the basic research level. It was mostly grants to universities looking at computational and experimental models and experimental data [for] reentry bodies, hypersonic cruise vehicle stuff, and the like." Very interesting, but much more theoretical than practical.

Walker had always admired DARPA's work from afar. "DARPA, for people in my field, was always sort of the crown jewel in places to work in S and T [science and technology]," Walker said. "I had always thought when I was at AFOSR and prior, 'Boy, it would be a great place to work.'" Walker had even run several programs for DARPA program managers while in the air force. "DARPA has a great reputation . . . in the science and technology community for getting things done," he explained to me later from his vantage point as a DARPA program manager. "I would say, the last five to ten years, this is the place that is doing most of the X-plane and technology demonstration work. A lot of that has dried up in the service laboratories. So, if you're an aero guy and you actually want to fly some things, DARPA is the place to be."

Walker joined DARPA as a program manager in 2003, for an initial three-year term, and began working on the details of what became DARPA's Falcon program. Falcon's stated application was primarily for a conventional strike missile that could be launched from a base in the United States and hit a target anywhere in the world in an hour. But the Holy Grail would be a successful demonstration of the technologies for doing much more.

"We really need to take an incremental flight demonstration approach," Walker told Tether, "where we can develop some technology, integrate it on a flight-test vehicle, fly it, learn, and incorporate that learning into the next vehicle, fly it, learn, and incorporate *that* into the next vehicle." Tether agreed, and the resulting program, as Walker later explained it to me, would use "a multi-flight vehicle demonstration approach." DARPA, under Walker's direction, would fund the development of not just one, but several vehicles, built one at a time over a period of years, each designed to prove out a different set of technologies that would ultimately go into the final, reusable technology demonstrator. "This is not like just flying four flights of the same vehicle," Walker told me. That approach—concentrating all the financial and political as well as technological risk into just one make-or-break vehicle—had proved NASP's undoing. Where the X-30 was to have been a large-scale vehicle, ready to be put to use immediately as a bomber or a transport, the cul-mination of Walker's Falcon program would be a small-scale unmanned vehicle, no bigger than a fighter plane, that would merely demonstrate the technologies needed for a larger vehicle without serving any intrinsic mission profile itself. "Then, at the end of the program," said Walker, "you could say, 'Wow, we've flown all the key enabling technologies using a series of different flight vehicles. We understand them now. Now the country can make an informed decision about whether we want to go forward and actually build and fly an operational system that incorpo-rates all these technologies.'" If the project lost political momentum or stumbled technically at any step of the way, it would still produce useful technologies. "We weren't going to wait until 2020 to produce some-thing," Walker told me. The plan called for enabling the development of working missiles within the decade, reusable unmanned vehicles the following decade, and full-scale transports by 2020 or 2025. That final step would, of course, depend on whether the country again wanted to make hypersonic transports a priority, but all the technologies would be in place to allow it to happen.

Walker studiously avoided any suggestion that he and his colleagues were directly working on the technologies for hypersonic civilian trans-ports. "Hypersonics tends to get oversold," he said. But he did acknowl-edge the obvious connection. "I think certainly if we do this work up front on the military side," he told me, "and improve the technologies for military use and operation use, those technologies could be applied to the commercial sector in terms of moving passengers much faster than we do today."

For the Falcon project, Walker and his contractors initially decided to build just three test vehicles. The propulsion system itself wasn't even to be part of the project, since NASA then had its own hypersonic engine project, the X-43. Instead, as he told me later, "We were going to work all the technologies except for air-breathing propulsion." Walker figured he and his contractors would have their hands full trying to develop structural materials that could survive sustained flight through the atmosphere at not just five, but six times the speed of sound. They would also work on navigation and guidance controls, and aerodynamics. All while NASA continued to perfect the technology needed for the engines. Or so they thought.

"These three vehicles were really glide vehicles," Walker told me later. "We were going to use them on rockets . . . get them up to speed and altitude, separate them from the rocket, and then have them glide back down to Earth, gathering data all the while." Each vehicle, called Hypersonic Test Vehicle (HTV) 1, 2, and 3, would glide better than the one before it, and be able to withstand more atmospheric heating, culminating in the third vehicle, which, unlike its predecessors, would fly more than once, to demonstrate the materials needed to fly the same vehicle repeatedly without it disintegrating under the punishment. During flight tests, the glide vehicles, boosted by rockets rather than air-breathing engines, would test the aerodynamics of the aeroshell shapes and the structural integrity of the carbon fiber–reinforced carbon composites of which they were made. "We're hoping that the tests will validate our models," explained Walker. "A lot of that has to do with the ablation and whether the leading edges on the aeroshell hold up like we think they will."

To get those unpowered test vehicles up to the speeds required to test their structural materials and guidance systems, Walker and DARPA funded a pair of conventional rocket projects under Falcon. The best known of these was the Falcon 1 rocket built by Space Exploration Technologies Corporation, or SpaceX, based in the Los Angeles area, with a launch site in the Kwajalein Atoll in the Pacific Ocean some 2,500 miles west of Hawaii. SpaceX was founded and is run by South African entrepreneur Elon Musk, who initially made his fortune in the heady days of the computer technology boom.

After co-founding the popular online payment service company PayPal and then selling it in 2002 for $1.5 billion, Musk founded SpaceX with the goal of undercutting the commercial satellite launch market by as much as one third, and using the profits to build the infrastructure needed to colonize Mars. "If we ultimately wish not to go in the direction

of the dinosaurs," Musk told me, "we will have to become a multiplan-
etary species." He was speaking of the asteroid strike theorized to have
wiped out the dinosaurs and which could occur again. "And the dinosaurs did not face the risk of self-annihilation that we do; all they could
do is try to eat each other one at a time."

Like XCOR, with its liquid oxygen pump, Musk and his team of
engineers took a very practical and well-considered approach toward realizing their lofty dreams. After Musk financed development of his rocket,
Walker paid for the first two flights of the SpaceX Falcon 1 satellite
launcher, with a view toward fostering a relatively inexpensive booster for
his planned glide vehicles. Falcon 1, which Musk named after the scrappy
Millennium Falcon starship of the *Star Wars* movies independently of the
DARPA program of the same name, failed in those first two attempts to
reach orbit. "The second launch I would say was about ninety percent successful," Walker told me, even though the vertically launched two-stage
rocket failed to send its dummy payload into orbit because sloshing fuel
in the tanks of the second stage unexpectedly threw off the guidance
system. "The engine worked great," said Walker. "It staged well." Falcon
1 ultimately made it to orbit on its fourth attempt, in 2008.

Boosted in part by the DARPA money, SpaceX went on to start
construction of its Falcon 9, powered by nine of the company's Merlin
engines, one of which powers the Falcon 1. Musk planned to send cargo
and crew into low Earth orbit with the Falcon 9 and, if all went well,
to service the International Space Station under NASA's Commercial
Orbital Transportation Services program after the space shuttle retired
in 2010.

SpaceX, by Walker's estimate, was well on its way toward revolutionizing access to space by making it much less expensive. "What we've
proved with SpaceX is you don't need a big team," Walker told me. A small
company and low overhead meant cheaper launches to space—under $10
million to launch a satellite, versus the $25 million and up charged by
the competition. "You can do things cheaper in this business if you're
willing to take a bit more risk," said Walker.

The other company building conventional rockets under the DARPA Falcon program was a startup working in Mojave, California, called Air-Launch LLC. That company's founder, Gary Hudson, had been trying to develop affordable access to space for years, most notably with another company with facilities in Mojave called Rotary Rocket, which tried unsuccessfully in the late 1990s to launch a single-stage-to-orbit craft that used a novel rocket-propelled helicopter system on top for landings. Several of the Rotary engineers stayed on in Mojave after the company went bust and founded XCOR Aerospace, whose liquid oxygen pump had led me to DARPA in the first place.

Now Hudson was at work on the AirLaunch system for DARPA and the air force. With no other investors to placate or other customers to please, the company kept a low profile, building and test-firing rockets in the desert, and attracting little notice with tests of its system for launching rockets out of the back of C-17 cargo airplanes. The plan did away with expensive and unwieldy first-stage rockets, allowing for much smaller rockets to launch the same size payload.

Like XCOR, AirLaunch sought to lower costs by doing away with costly turbo pumps. Instead, the kerosene fuel for the company's rocket engines was pressurized right in the tanks and fed through high-pressure lines to the engine for combustion. Not only was the system cheaper and quicker to launch, but it had the additional benefit of enabling a launch from anywhere a C-17 could fly, not just from a dedicated missile base or spaceport. Like the engineers at XCOR and SpaceX, Hudson hoped to work his way up to manned flights to low Earth orbit and beyond. "It has a lot of interesting capability," Walker told me of the AirLaunch rockets he was funding. "An AirLaunch capability that is simple and quick and gets you away from the [traditional launch pad] is a very useful thing to have."

Useful not just for boosting hypersonic air-breathing test vehicles up to operational speed, but also for the armed force's desire for what came to be known as Operationally Responsive Space. The air force, for one, wanted small, inexpensive satellite launchers capable of putting up spysats and communications satellites practically at a moment's notice to support operations anywhere in the world. The rockets of both SpaceX and AirLaunch might just fit the bill, and DARPA was only too happy to foster development of that capability as an offshoot of its hypersonics program.

Meanwhile, work on the hypersonic test vehicles had progressed faster than the conventional rocket boosters originally conceived to boost them up to test speeds. "So we've gone out and procured the Minotaur launch vehicles," Walker told me, "because those exist and have flown before." The Minotaur rockets were built around motors cannibalized from decommissioned ICBMs by a company called Orbital Sciences. A decade before, the company had built an air-launched satellite booster for DARPA called *Pegasus*, which gave DARPA and the air force part of the answer to cheap access to space, but not all of it. "What we were really looking for was something even more responsive than *Pegasus* and actually more affordable," Walker told me of the decision to fund SpaceX and AirLaunch.

However, fate, or rather President George W. Bush, intervened in Walker's plans even as his Falcon program got started. In January 2004, Bush announced his Vision for Space Exploration, which called for NASA to drop everything and focus all efforts on returning humans to the moon, and from there on to Mars. The NASA X-43 hypersonics program that Walker had been counting on to do his propulsion research got canceled. The X-43 went out in a blaze of glory later in 2004 when the project's one-time-use hypersonic test engine launched from a B-52 bomber and broke the air-breathing speed record by going faster than Mach 6. In spite of the X-43's success, however, NASA would get no further funding for the X-43, and Walker and Tether had to switch gears.

"Dr. Tether exercised one of the options on the contract," explained Walker, "which was to do an air-breathing propulsion flow path." In other words, DARPA launched its own hypersonics propulsion program. Walker put his contractors to work on the necessary components of a working hypersonic air-breathing engine as part of a program within

Falcon called Falcon Combined-Cycle Engine Technology, or FaCET. With no moving parts, a hypersonic engine is all about flow path, and DARPA got to work developing the major components of that flow path—the inlet, where the onrushing supersonic air enters the engine; the combustor, where the air meets the engine's fuel and, with any luck, ignites; and the nozzle, where the resulting gases escape to thrust the vehicle forward at, it is hoped, hypersonic speed. By 2006, Walker and his colleagues were able to pitch a not-entirely-unexpected idea to Tether. "How about if we modified HTV-three," Walker asked Tether, "to not just be a reusable glide vehicle, but let's put an engine on it and test the propulsion system as well?"

Tether "did one better on us," as Walker put it. "It would be really nice to really prove hypersonic airplanes," Tether told Walker. "So rather than using a rocket to boost this thing, why don't we design it so it takes off like an airplane?" Tether figured if the demonstration vehicle could take off on its own, just as an operational vehicle would, it could demonstrate the greatest possible range of uses for the technology. Once airborne, the test ship would kick itself up to hypersonic speed, and then slow down for a landing under its own steam. And of course, it went almost without saying that it would have to do this repeatedly to fully prove out the structural materials and other technologies that would go into a full-scale vehicle.

The ability for a hypersonic jet to take off from a runway under its own power presented a major technological challenge. "People have flown hypersonic demonstrators in the past," Walker explained to me "but generally they have used a rocket to boost the demonstrator up to a certain Mach number, and then the supersonic scramjet takes over, usually at about Mach five or six." The whole idea of building a hypersonic jet in the first place is to do away with rockets and their inefficiencies, including the need to haul around more than their own weight in oxidizer with them wherever they go. Thus, "the DARPA challenge," Walker said in summing up the crux of his program as it had evolved by the time we spoke in 2008, "is to build this combined cycle propulsion system that can take off like an aircraft, get up to hypersonic speeds of five to six times the speed of sound, and then come back down and do it again." That, in fact, is the "DARPA-*hard*" challenge, as program managers referred to the perhaps impossible tasks they tackled in their work. Building a hypersonic airplane was now possible, if only just barely. But no one had yet succeeded in making the whole system an air-breather at every phase of its flight.

With little fanfare, the originally planned HTV-3 glide vehicle became the HTV-3X concept design for a powered vehicle. It was called a concept design because it was to be just that: a concept rather than an actual working vehicle. "Under Falcon, we don't have the funds" to fly a working vehicle, Walker explained. "HTV-three was going to fly on a booster, and we modified it to include a propulsion system. We said we would take it to concept design and then see." Although the HTV-3 wouldn't actually fly, the technologies would nevertheless go far beyond mere thought experiments. "It's not just paper," said Walker. "There is a lot involved. . . . A lot of the propulsion testing. We've done wind tunnel tests. We've done materials tests. We built a fuel-cooled liner. There are a lot of risk-reduction activities that go along with the concept design."

Risk-reduction activities such as the wind tunnel testing I went to NASA Langley to see. While the X-51A's engine undergoing testing at the time of my visit would not be a combined cycle engine, it would feature many of the same technologies, including the fuel-cooled design. True to his mission to realize the dream of hypersonic flight in a step-by-step process, Walker was already overseeing a DARPA and Office of Naval Research collaboration known as the Hypersonics Flight Demonstration, or HyFly. The plan was to launch a rocket-boosted scramjet from an F-15 fighter plane off the coast of California and get it to fly at Mach 6 for at least 100 seconds. Not coincidentally, those specs—small enough to launch from an F-15 and sustained Mach 6 flight for more than a minute and half—would make such a vehicle an ideal hypersonic missile, the first practical application planned for hypersonics. If successful, HyFly would push the envelope for hypersonic jets far beyond the current state of the art. But to demonstrate the ability to fly anything other than a warhead, a scramjet engine would have to be capable of a lot more than one hundred seconds of flight, and for multiple flights rather than just the single-shot missions of HyFly and every flying scramjet before it. Achieving those goals was the mission of the X-51A Flight Test Program.

Already the program had achieved record-setting performance in the lab. A test engine had run at Langley for a combined total of almost 20 minutes during separate tests. "We tested it at Mach 4.6, 5.0, and 6.5," Curtis Berger, manager at Pratt and Whitney Rocketdyne, the test engine's builder, told me. "The amount of time that this thing was actually running and creating thrust was just about 17.8 minutes. Over 17 minutes of time on this engine," he repeated for emphasis. "That's a lot of time for a scramjet engine." And this was putting it mildly.

Second by second, the engineers from Pratt and Whitney Rocketdyne had been pushing their test engines beyond the limits of endurance for any other scramjet engine. The engines' steel-nickel-alloy frames, coated with carbon mesh for heat resistance on the leading edges, were designed to withstand temperatures soaring to 2,100 degrees Fahrenheit. But even these materials weren't enough protection for the little engines that could, so they were also actively cooled by their own fuel—the same JP-7 jet fuel that had been used by the SR-71 Blackbird. Using more or less ordinary jet fuel instead of, say, the more energetic—but also more finicky—liquid hydrogen used by NASA's X-43 scramjet was an important aspect of the program, since it would allow operational engines of the same basic design to use existing jet fuel distribution and storage systems.

Once the engineers wrung all the test data they could from the engine in the test chamber, they'd build the X-51A flight-test engine and, if all went well, use it to power the X-51A unmanned test vehicle for an unprecedented five minutes of flight. Just to be clear, Berger told me, "The five minutes of flight we're talking about is not limited by the propulsion system. That's just how much gas we have in the tank." This major breakthrough was just around the corner, and it would be a big step toward building practical scramjet-powered vehicles.

The sense of excitement during my visit to Langley was palpable, but muted. The men working on the project had spent big chunks of their careers on hypersonics research, and although they were closer than ever now to realizing the dream of sustained flight by an air-breathing engine at speeds faster than Mach 5 and even surpassing Mach 6, they had experienced plenty of disappointment along the way. AFRL program manager Charlie Brink, who had joined the NASP project right out of college as a young air force lieutenant in 1984, described himself and his cautiously optimistic colleagues as having a "NASP hangover." Just because things were going well now didn't mean the field wouldn't have the rug pulled out from under it again.

The engine in the high-temperature tunnel at the time of my Langley visit was the last of a line of test models needed to prove out the technologies before an actual test flight. If all went well here, the next stop was 49,500 feet over the Pacific Ocean, when the engine's progeny would be dropped from a B-52 bomber to take flight on its own for the first time. That final phase of the X-51A program couldn't launch, however, until the following year—after a new presidential administration took office, one that might have entirely different priorities than the previous one. Perhaps the AFRL's Brink summed up his and his colleagues' attitude best when I asked him how he coped with the ups and downs of scramjet research. He responded by hunching over the table before him and making a scribbling motion. "Just keep your head down and color," he said.

After a briefing, the engineers took me into the high-temperature tunnel's control room. Its heavy steel doors reminded me of those of a bank vault. Anyone in the tunnel when it fired would instantly be incinerated in the hypersonic blast of hot gas and air that roared through during the tests. The tunnel's designers had put elaborate safeguards in place to ensure that no one would be caught inside. The heavy, locked doors were one of them. As a further safety measure, the technicians who worked in the tunnel to prepare the air- and spacecraft models for testing carried the keys needed to activate the tunnel with them. No one could start the tests until the technicians returned and turned their keys in the banks of control knobs and levers on the walls.

The test operators sat at chunky cold war–era control consoles that had had their original CRTs yanked out and replaced with LCD monitors. More LCD monitors graced the walls above the control panels there. Since a test wasn't actually scheduled for that day, we were also able to go into the high-temperature tunnel itself. We reached the interior of the test chamber via a short metal stairway, and I found myself in an eight-by-twelve-foot spherical enclosure that looked like an overgrown bathysphere.

The Eight-Foot High-Temperature Tunnel, or HTT, had been designed at the close of the 1950s and finished in the mid-1960s to enable testing of designs for spacecraft reentering the atmosphere. The tunnel became, and still is, the biggest hypersonic test tunnel in the United States. No mere fan blades, no matter how large, could generate the kind of force needed to replicate the conditions a speeding air- or spacecraft would encounter traveling at hypersonic speed through the upper atmosphere. Successive tanks of compressed air, each boosting the force of the one before it, could do the trick for a small-scale test chamber, like others at Langley, but not on a scale large enough to test vehicles of any significant size—at least not without sucking down the power output of a major electric generating plant. So the tunnel's designers devised what amounted to a giant rocket engine to get the job done. Instead of spinning big fan blades, the HTT ignited an eight-foot-high jet of methane gas to generate the heat and dynamic pressure of hypersonic flight.

Initially designed for rockets, the place got an upgrade in the 1990s to accommodate air-breathing engines. The methane jet burned up all the oxygen available to it from Langley's array of compressed air storage tanks, known as the tank farm, leaving nothing for anything else to breathe in the test chamber. Not a problem for the uses for which the facility was designed—testing thermal tiles and other heat protection, as well as aerodynamic forms for capsules and other spacecraft—but a disaster for a jet engine that might actually try to sustain combustion at hypersonic speeds. So, in support of the NASP program, the HTT got a liquid oxygen injector to add back just the right quantity of ambient oxygen a vehicle in flight would be able to suck in at fifty thousand feet or so.

Inside the test chamber, we all gathered around the test engine on its copper pedestal. The engine had been installed upside down to enable all the internal plumbing, fuel pump, and tanks to be housed inside the pedestal. Once the tunnel's methane jet was ignited and the heat and airspeed conditions brought up to full force, the pedestal was raised from the floor of the test chamber and into the flow of hot gases. Test over, the pedestal would descend back into the floor of the chamber. Waiting until the test chamber was brought up to full force before thrusting the engine into it, and then lowering it out of the maelstrom before the tunnel wound down, saved the test engine from needless wear and tear when it wouldn't actually be collecting data.

The snout of the engine itself pointed into the force of the blast, which would emerge from the yawning exit nozzle of the HTT. Looking down the business end of the HTT, easily large enough to accommodate two men walking two abreast, I felt a new appreciation for the tunnel's safety measures. The rectangular exhaust nozzle of the test engine pointed out the other end of the tunnel, where its combustion products would be mixed with those of the tunnel itself and blasted out (along with any pieces of the engine that might fail during the test) into an expanse of uninhabited swamp.

Fuel flowed through the walls of the engine to cool it while heating up in preparation for ignition in the hot supersonic air already rushing down the engine's throat. The now-hot fuel, thus "conditioned," would be primed for wicking yet more heat away from the engine, thus enabling it to go faster without danger of it disintegrating in the terrific heat. At that point a supercritical liquid, the fuel would flow through valves on the top of the engine for injection into the engine as a vapor, which would pack fully 10 percent more energy by volume than the liquid form of the fuel, boosting the engine's performance even further. An ethylene torch would ignite the vaporized fuel to light the match in the hurricane. It would be the job of an off-the-shelf jet engine control computer to precisely modulate the fuel injection and keep that match lit. As far as what the engine actually looked like inside—that all-important interior geometry that made hypersonic flight possible—that was a secret that I wouldn't be permitted to see. Nor was I allowed to bring my camera into the building.

During an actual test run in the wind tunnel, the scene was reminiscent of the countdown for a rocket launch. Only, instead of hunkering down in a bunker while monitoring a rocket outside, the engineers for this test sat practically in the same room as the thing. During a test before my visit, the engineers watched the readouts and dials in front of them, as well as the video monitors showing the yawning tunnel on the other side of the wall, for any sign of trouble as valves opened and pumps went to work in preparation for igniting the tightly controlled inferno.

The tunnel was quiet now, but that was about to change.

The countdown proceeded without incident, and on "zero," a jet of blue methane fire lit up the inside of the tunnel and blasted down its entire twelve feet in length. Insulated from the blast in the control room, the engineers nevertheless could hear the jet as a low rumble.

Again they checked their instruments as the methane jet roared to full power. Satisfied that all was well, the test director gave the command they had all been waiting for: "Okay to inject."

The pedestal rose from the floor of the tunnel. It had been constructed of copper, a metal that readily dissipates heat, to enable it to survive the blast. The sharply raked lines of the test engine, perched in its place of honor atop the pedestal, screamed speed.

"AOA modulating," the test director informed the group as the pedestal brought the engine into position and began to settle it into place. Then, a moment later, "Model on center line," he said as the pedestal locked the engine into place. Finally: "We are in ignition."

The jet of the test tunnel burned at more than two thousand degrees

Fahrenheit. But the fire that erupted from the back of the engine was even hotter. It grew brighter as the engineers throttled the engine up. Sixty anxious seconds later, the engineers shut the engine off and lowered it back through the floor of the test chamber and out of the path of the tunnel's methane jet. The test was an unqualified success.

If all went according to plan, the X-51A would fly in late 2009, dropping off from a B-52 bomber out of Edwards Air Force Base in California. Over the Pacific at 45,000 feet, the vehicle would fire a solid-fuel rocket to reach operating speed at 60,000 feet and 4.5 times the speed of sound before dropping the rocket and lighting up to push past Mach 6 and 80,000 feet in altitude. After transmitting its test data to the engineers back at Edwards, it would dive into the ocean in a controlled crash, its fuel spent. Throughout its flight, the X-51A, whose airframe and integrated control surfaces would be built by Boeing, would remain under the control of the flight-test engineers. "This is an airplane," Pratt and Whitney Rocketdyne's Berger told me, to emphasize its status as a practical flight demonstrator rather than a mere lab experiment or a missile test. "This is really beyond something you might do for a weapon application. The whole idea is to prove the practicality of a free-flying, scalable, scramjet-powered vehicle." The AFRL's Charlie Brink, for his part, as he only half-joked to me, would probably be locked in a men's room at Edwards throwing up, as the focus of all his hard work over the last few years reached its make-or-break point.

Assuming a successful flight of the X-51A, it would take only one more step to complete the pieces of technology necessary to breathe life into the dream voiced by President Reagan—minus the maximum speed of Mach 25.

DARPA codified that final step with its Blackswift program in 2008. Blackswift was envisioned by Walker as a small-scale demonstrator for all the technologies needed for a full-up hypersonic transport. "We actually have a competitive solicitation out right now on the street," Walker told me in early 2008. He invited the engineers working on the Falcon program and other potential contractors to propose how they would fit all the different pieces of the technological puzzle together into the world's first reusable air-breathing hypersonic vehicle cable of sustained flight on its own, including take-offs and landings. "I really can't say much more on it," Walker told me. "We need to wait and see what kind of proposals come back."

By this time, he was in for the long haul. To follow an initial three-year term, Tether signed Walker up for another two years as a program manager, for a total of five. Like those of his colleagues, Walker's badge displayed the month and year in which he had to leave the agency. When I spoke with him in March 2008, it said his last day would be the following month. But by then he had been promoted to the deputy directorship of the Tactical Technology Office. "Generally when that happens, your time starts over," Walker told me. "I was thinking on the way into work that I really need to talk to [Tether] about that." Sure enough, Walker was still at the agency when I next spoke with him at DARPA headquarters in June 2008. By then he was going on six years with DARPA. "So my time is getting short." Nevertheless, it seemed he would get to stay on long enough to see the dream of reusable hypersonic flight with an air-breathing engine realized. "People like to talk about 'DARPA-hard,'" Walker said. "This is *really* DARPA-hard." Sustained flight at five, six, or even seven times the speed of sound had never been achieved, but now it seemed within reach. "If we do everything we say we're going to do in Falcon," said Walker, "[it] will enable the country in five or six years from now to have all the data it needs to say, 'Hey, we want to go build a hypersonic airplane for military or commercial use.'"

Conceptual artwork of the Blackswift showed an unmanned fighter jet–size craft with a sharklike prow, a slim dark gray fuselage, and ovoid engine inlets. If all went according to plan, by 2012 this ship would take off under remote control from a runway at Edwards Air Force Base under its own jet power, with a chase jet in pursuit. Blackswift would lose the chase jet not long after reaching Mach 1, still traveling under conventional jet power. When it got to Mach 2, it would undergo a remarkable transformation that would allow it to bypass the limitations of its jet turbines. Inside the jet's inlet nozzles, flaps would shuttle the airflow away from the turbines and straight out through the engine outlets, turning a conventional jet (albeit a very fast one) into an even faster scramjet. After sustained flight at Mach 6, the airplane would shut down the scramjets, coast back down to subsonic speed, and spin up the turbines once again for a powered landing at Edwards.

Unfortunately it was not to be—at least not on the timetable Walker and Tether had hoped for. In September 2008, Congress cut the funding DARPA and the air force had requested for the collaborative Blackswift, from the necessary $120 million ($70 million from DARPA and $50 million from the air force) to just $10 million. Blackswift had no chance of getting off the ground with that budget, and Walker had no choice but to withdraw his request for proposals. But he refused to give up hope. "It was a good idea," he said of Blackswift at the time, "and good ideas have a way of coming back and getting done eventually."

Thus far the United States and its DOD-funded research appeared ahead of the curve on hypersonics, but other nations were working hard to catch up. At the American Institute of Aeronautics and Astronautics hypersonics conferences, held in the United States, Chinese researchers had been slowly leaking details of their own scramjet research programs. This work includes constructing a hypersonic wind tunnel in the Chinese capital and designing rocket-scramjet combined-cycle engines. One American scramjet researcher told me he steers clear of the conferences because the Chinese have the disconcerting habit of cornering their American counterparts in the convention hotels and plugging them for all the information they can possibly get from them. "I would bet that they have a serious research program under way that has a lot more going on than just the few papers they have issued at this forum," *Aviation Week & Space Technology* editor Craig Covault told me after the AIAA Joint Propulsion Conference in 2007.

Not all hypersonics researchers are focused solely on military applications. In particular, the work of European Union researchers give a hint of what a hypersonic civilian transport, an airliner, might look like.

Even as DARPA's Steven Walker and the AFRL's Brink and their contractors keep their heads down and keep coloring, the European researchers *are* directly investigating the commercial applications for hypersonic air-breathing engines as civilian transports.

The EU researchers have nothing like the resources of DARPA and the U.S. Air Force, and their investigations into hypersonic passenger jets are mostly relegated to the realm of studies rather than the experimental research of their American, Australian, and Chinese counterparts. Nevertheless their work presents an intriguing look at what might well be the future of high-speed flight.

As part of its Long-Term Advanced Propulsion Concepts and Technologies, or LAPCAT, program, the European Union commissioned the UK-based company Reaction Engines to study the problem of creating an aircraft capable of reaching Sydney from Brussels within four hours. That's halfway around the world at Mach 5, or faster. The LAPCAT project was set up to produce studies only, not working hardware, but it was enough to lay the conceptual groundwork for the construction of actual vehicles in the future. Reaction Engines's concept, called the A2, has seating for three hundred passengers.

The company's technical director, Richard Varvill, believes any such concept is viable only if it overcomes the limitations that rendered the world's only supersonic passenger transport, the Concorde, little more than an expensive novelty. "The range was inadequate to do transpacific routes, which is where a lot of the potential market is thought to be for a supersonic transport," Varvill told me of the Concorde. To be truly viable as a means of transporting large numbers of people affordably, the next supersonic (or hypersonic) transport would have to be able to reach across the Pacific Ocean to link the United States and the burgeoning Asian markets.

The other big problem with the Concorde, as Varvill sees it, was that the ship's engines were so perfectly tuned to its Mach-2 cruising speed that they couldn't run efficiently at subsonic speeds. That wouldn't have been such a problem had the airplane been permitted to maintain its optimum speed for most of its flight, but after the Concorde was introduced in 1969, the sonic booms it produced created such a nuisance for people on the ground that new regulations forbade it from cruising past the speed of sound when traveling over land. Because the airplane got such bad gas mileage when traveling under the speed of sound, the airlines that owned it suddenly found themselves unable to operate it economically on overland routes. That plus its range limitation meant that the Concorde was relegated for the last years of its operational life to just one route—the U.S. East Coast to London and Paris. Speaking of the A2 design, Varvill says, "The engine has two modes, because we're very conscious of the Concorde experience."

Two modes were just what Walker wanted for the HTV-3 and Blackswift: a turbine-powered mode for operating efficiently at subsonic speed as well as at low Mach numbers, and a scramjet component for high Mach numbers. In the A2's case, not only were dual modes essential for providing a means to get the ship up to the supersonic speeds its precooled hypersonic jets needed to work, but they would give the ship the ability to operate affordably at the conventional airliner speeds necessary for avoiding sonic booms, and that in turn would give the ship the overland routes denied the Concorde.

Like the X-51A engine that I saw in the test chamber at NASA Langley, A2's engines would most likely use their own fuel—in this case, liquid hydrogen—to keep themselves cool. To avoid overheating from atmospheric friction and suffering structural failure, the ship would also circulate the fuel throughout its body, which would require it to be built without windows. Ordinary aircraft windows also lack the structural strength to survive prolonged exposure to Mach-5-plus flight through the atmosphere. Thus concept art of the A2 shows a rapier-prowed, tailless fuselage unbroken by windows. External cameras might feed outside views to passengers via seat-back screens. Anyone who objected to the lack of windows would likely be glad to trade such a minor inconvenience for the luxury of shaving a twenty-two-hour flight down to just four.

Besides packing more energy per unit of fuel, hydrogen gives the A2 the additional advantage of producing no environmentally damaging carbon emissions—just water vapor as the product of hydrogen combining with oxygen during the combustion process. Varvill is careful not to try to sell his company's concept based on the supposed environmental merits, however, believing that the much-vaunted hydrogen economy is even farther from coming to fruition than his aircraft design. Hydrogen will, for the foreseeable future, Varvill believes, continue to be produced in large, commercially viable quantities only by using carbon-dirty processes such as extraction from fossil fuels.

More than twenty years after Reagan's premature announcement of the feasibility of hypersonic airplanes, it seems the technology is at last reaching fruition. Proponents of the technology at DARPA and elsewhere bided their time and now it looks as if they are close to reaping the rewards. APL's Van Wie is philosophical about the decades it has taken for this to happen. "Advanced propulsion technology has a development timescale that appears to be on the order of decades," he told me. He pointed out that more than forty years had passed from Konstantin Tsiolkovsky's seminal 1903 paper—in which the Russian scientist laid the theoretical groundwork of modern rocketry—to Germany's V2 rockets, built at the end of World War II. Considering a 1960 conference at which the basic theoretical framework for scramjets was laid out as playing the same role for hypersonics as Tsiolkovsky's paper did for rockets, Van Wei figures hypersonics are about due for their place in the sun. "So if you look at that—1960 to now, forty-seven years or so—it's kind of on the same timescale to see this roll out." DARPA's Steven Walker concurs, but cautiously. "We are in a period here where we have an opportunity to develop and demonstrate . . . all the technologies that would be required to fly a hypersonic airplane," he told me in his characteristically understated way.

A lot has to go right, not only in the lab and in flight tests, but on the political front as well, for the money to continue to flow into hypersonic research projects. Then and only then might those holding the pursestrings be in a position to evaluate whether the technology is worth bringing to full realization in operational vehicles. "I think what the country needs to remain committed to," Steven Walker told me, "is at least demonstrating all the technologies required—propulsion, aero, materials, how you navigate and guide the vehicle over long hypersonic flights. And then sometime in the next decade, after that's been done, decide, either for military or commercial use, is it worth the investment? What do we get out of it? How much is it going to cost? A lot of those answers we won't know until we actually fly some of the technologies." Getting as many of those answers as possible, through development and testing of actual flight hardware, was what Walker had committed himself to doing during the rest of his tenure at DARPA.

"We often say," Walker said of himself and his colleagues at DARPA, "that we're great at taking the technology excuse off the table to pursue new capabilities." If DARPA's managers and their contractors succeed in this instance, hypersonic airplanes might one day take their place among the many other groundbreaking technologies funded by DARPA that have quietly made themselves indispensable to the modern world.

POWER TO THE PEOPLE

"OUR FOCUS TODAY is on 'tactical energy independence,'" Douglas Kirkpatrick told the crowd during his 2007 DARPATech presentation, "but the strategic implications of these technologies are hard to miss." He had just finished laying out how the Strategic Technology Office was working to revolutionize energy production for the U.S. military during conflicts, including with the most efficient solar cells ever made and jet fuel made from vegetable oil, and now he was summing up. "Writ large," he said, and in his TelePrompTer notes, these words were italicized, "these new technologies would transform the geopolitical environment into one in which energy was completely removed as a source of conflict."

In other words, the new energy production technologies Kirkpatrick's office was working on could not only unleash combat soldiers from the long supply chain of energy—most notably fuel—that they have always been dependent on, but could also remove the need for them to go into combat in the first place.

That Kirkpatrick's words drew no controversy at the conference was an indication of just how obviously U.S. dependence on foreign oil, particularly from the Middle East, was hurting every sector of American life along with the military. "We are a strategic agency," Kirkpatrick told me after the conference, "and I sit in . . . the Strategic Technology Office, and the strategic implications of very high efficiency solar should be implicitly obvious, and they're not lost on us." He left me to connect the dots to the current war in Iraq, but clearly energy—namely, petroleum—played a starring role in that conflict.

"Ultimately," he repeated to me, "the strategic goal is to remove energy as a source of conflict worldwide. What I was most gratified by is that I was allowed to say that publicly. That shouldn't be lost on people—that DARPA at this point in time is engaged in the strategic energy fight. That is," Kirkpatrick reiterated, "perhaps the preeminent agent of change technologically within the U.S. government . . . is now in the fight for energy technology. That's a big deal."

Leaving aside the greater implications of his work, Kirkpatrick's immediate aim was to change the way the United States headed into battle. Energy, he pointed out in his DARPATech talk, is the single most important factor in waging war. Not just now, but in every conflict in history, he said, "energy has been *the* limiting factor in all military operations." The American convoys then rolling through Iraq presented some of the most attractive targets for the enemy, and most of their load was made up of fuel and batteries, he said. The energy load took its toll on the individual soldier in other ways as well: fifteen to twenty pounds of the typical one-hundred-pound load of supplies each soldier had to hump through enemy territory was made up of batteries. "At DARPA," said Kirkpatrick, "we see a situation like this and we ask ourselves: 'What if we could liberate the war fighter from the endless energy logistics train?'"

It was a bold idea that would require a radically different approach to energy use. Instead of a centrally distributed energy system—i.e., trucks and supplies moving from one location to soldiers and installations in the field—Kirkpatrick's vision called for soldiers and bases in the field to generate their own power so they wouldn't have to carry it with them or depend on resupply. "We're talking complete self-sufficiency," he said. "We need infinite sustainability, the ability to draw all our energy needs from the environment around us." Any technologies that made this possible, if successfully developed, could also upend the civilian world's approach to power generation, something Kirkpatrick knew full well.

He used the term "energy independence" in his DARPATech presentation, because that was the commonly accepted phrase for reducing America's dependence on foreign oil. He later admitted to me that he really didn't like that term. In fact, he said he hated it, "because it smacks of focusing inward as opposed to focusing on the world as a whole. When I say energy independence, what I mean is energy independence from petroleum." By then he was talking both about the military and the civilian world. He preferred to use "energy source diversity," to convey the notion that the nation should spread its energy risk by drawing from many different sources. As Kirkpatrick put it, "Why would you want to take your obedience to the great master petroleum and switch that for some obedience to some other thing?" He had made it his mission at DARPA to help free the United States from what he called the tyranny of oil.

Douglas Kirkpatrick majored in physics and mathematics at the College of William and Mary, in Williamsburg, Virginia. He earned his Ph.D. in plasma physics from MIT in early 1988, and went to work for the Naval Research Laboratory, or NRL, in Washington while teaching at the University of Maryland. He couldn't make up his mind between remaining in academia and going to work as a scientist for a corporation. He was torn, as many DARPA program managers have been, between the choice of pursuing cutting-edge research that may never lead to practical application and working on actual applications that didn't push the envelope very far.

In the summer of 1988 he got his first chance to work for DARPA, as a contractor, and he liked what he learned about the agency. It remained a constant presence in his career most of the next ten years, following him to a job at defense contractor Science Applications International Corporation, or SAIC. Finally, though, he felt he had to leave the defense industry. "The reason I left was because I very much wanted to make things, and I was getting more than a little bit frustrated that we would get up to the point of having a research success and then it would have to go someplace else to be turned into reality." He wanted to be in on both sides of the research-and-development equation—not only coming up with new ideas for applications, but helping to build them as well.

He found such an opportunity at a company called Fusion Lighting, where he spent the next five years developing novel ways to light industrial installations. There, he said, "I learned a great deal in terms of how you take technology and turn it into a product." Trouble was, any technology sufficiently advanced to engage Kirkpatrick's interest came with such heavy start-up costs that any company developing it, particularly a fairly small one such as Fusion Lighting, was likely to go bust before getting its product to market in sufficient numbers to realize a return on its investment. Kirkpatrick saw the writing on the wall at Fusion Lighting and decided to jump ship before the inevitable collapse occurred.

DARPA managers had had their eye on Kirkpatrick since his SAIC days, and it wasn't hard for him to wrangle an interview with Tony Tether in 2002. Tether took Kirkpatrick on initially to run a program called High Efficiency Distributed Lighting, or HEDLight, which created a centrally distributed system for lighting a big military installation such as an aircraft carrier, much the way an electrical generator distributes electricity through a building or ship. The idea was that the lighting system could be centrally protected in the bowels of the ship and reduce maintenance headaches caused by hard-to-reach light bulbs—by piping light around in fiber-optic cables.

Although the problems of energy distribution and efficiency were on Kirkpatrick's mind when he came to DARPA, solar power and biofuels didn't occur to him as obvious solutions to some of the major problems facing the armed forces until he began to interact with his new colleagues. "Energy was on my mind from the standpoint of lighting," he told me later. "I was always looking at energy efficiency, and I looked at photovoltaics, and I looked at biofuels, but I didn't really perceive that I was in a unique position to drive those things." It was through conversations in the halls and offices of DARPA that he began to put the ideas together for what may just turn out to be among DARPA's most influential projects yet. Kirkpatrick's real talent, he realized then, was as what he later termed a harvester. "It's pretty easy for me to see value and application potential in basic research work that others are doing, and ask the question 'Have you thought about this and have you thought about that?' This place is packed with brilliant people with just innumerable ideas."

Some of those ideas led to Kirkpatrick's Very High Efficiency Solar Cells program. He had been discussing recent breakthroughs in creating nanoscale (that is, molecular-size) structures with Defense Sciences Office program manager Eric Eisenstadt, who had funded the advances. The two of them hashed out some ideas for putting those structures to use, including in batteries, advanced optics, and solar power. They brought those ideas to Tether, who, as Kirkpatrick later recalled, immediately saw that the biggest potential was in solar power. "If you think you can do what you're talking about doing in solar," Tether told the other two men, "then that is far and away the biggest of anything you're talking about here." Solar cells built of nanoscale layers—with each layer precisely tuned to absorb a different frequency of solar energy—could theoretically generate more electricity from a given surface area than the broadly tuned silicon commonly used in solar cells, and Tether was eager to see just how far theory could be pushed into practice.

Once he'd formed the basic ideas behind what became the Very High Efficiency Solar Cells program, Kirkpatrick presented them at an Institute of Electrical and Electronics Engineers (IEEE) conference in Hawaii in May 2006. As Kirkpatrick told me later, he was shocked at the reception he got. And not in a good way. Seems the program's goal of converting fully half of the solar energy striking a solar cell into electricity was much, much too ambitious, as far as the attending engineers and scientists were concerned. At the time, even experimental solar cells could barely crack 20 percent efficiency. "We were vilified," Kirkpatrick told me, "because we were taking the best and brightest from the solar world and setting them off on this absolutely hopeless task, and all we were going to do was to prove that solar was not important." Far from being recognized as fostering innovations in the field, "we were the enemies of the solar world because we were going to show how useless it was." The problem, as the other engineers saw it, was that DARPA, through Kirkpatrick, was bent on setting up the best minds in the field for failure, and failure would make the whole field of solar power research look bad.

To Kirkpatrick, this was a perfect example of what he saw as the key difference between DARPA and the rest of the science and technology community—namely, academic and industrial researchers. The university researchers were rewarded for making small incremental improvements to existing technologies because they had to keep publishing their results. And those results had better be good if the researchers intended to get tenure. To put it another way, the researchers were evaluated based on what they published, and not on whether they produced solutions to problems in the real world. "You need, what, four, five, six publications a year to get tenure," said Kirkpatrick. "You can't afford to have failures, because you can't publish failure." On the other hand, industrial researchers were *too* focused on return on investment, which made them equally unwilling to take chances on potential breakthrough technologies. Academic and industrial research scientists, to Kirkpatrick's way of thinking, were thus united by an aversion to failure that hindered their ability to make real breakthroughs.

Hence the role of DARPA in pushing not just incremental innovation, but big leaps forward. Only the government could spend the kind of money needed to pull it off. But it wasn't just money that was needed, said Kirkpatrick, it was properly focused money. "DOE [Department of Energy] spent $150 million on solar, but they spent it across thirty or forty contractors. We spent $150 million on solar, and we put it all on one contractor." That "is a fundamentally different approach."

Not that Kirkpatrick was anxious to put down the Department of Energy. In setting up his solar power program, he worked closely with Craig Cornelius, his counterpart at the DOE, comparing notes on the most promising players in the area of solar power research. The DOE's focus was on funding scientific research that may not be close to application, while DARPA's current emphasis was on developing practical applications in a short time. "Craig would come to me," Kirkpatrick told me, "and say, 'These people are getting to a breaking point. They're really ready to go, but there is no way I'm going to be able to get the resources in DOE to push them along.' Or I would see something and say, 'Craig, I think this is close, but these guys just don't have the horsepower to play in our league. They need a little bit of nurturing.'"

Nurturing, Kirkpatrick explained to me, was the very antithesis of DARPA's approach. "DARPA is, 'Get on the train, suit up, and let's go play.' And some folks need a little bit more nurturing before that can happen. You can ask my wife. I'm not the nurturing type."

Although Kirkpatrick saw the possible commercial applications for his Very High Efficiency Solar Cells as a given, he was quick to emphasize when speaking with me the very specific purpose of the program: to reduce the number of batteries that soldiers would have to hump into the field. Extremely efficient solar cells could do that by generating the same amount of power as standard solar cells of much larger size. The more efficient cells could conceivably be made small enough to incorporate into the housings of electronic gear such as radios and GPS units, which would recharge every time they saw the light of day. Soldiers in camp could simply leave the devices out in the sun between missions to charge them up, instead of swapping out batteries. Not having to carry batteries would cut each soldier's load by a fifth, and would let troops move faster and go farther. Perhaps even more important, it would also reduce the amount of supplies trailing those soldiers by 10 to 15 percent. And fewer logistics supplies would mean fewer trucks trundling across dangerous territory behind advancing troops.

It wasn't enough for Kirkpatrick simply to have his contractors demonstrate 50 percent efficiency in a solar cell, which was in itself no mean feat. That represented a 250 percent boost in the energy-gathering ability of the best solar cells on the market and their paltry 20 percent efficiencies. No, the program would also have to demonstrate the means by which the cells could be manufactured in great enough quantities, and quickly enough, to supply troops with solar cells suitable for powering all of their electronic gear. Kirkpatrick used the rate at which newspapers are manufactured, a square meter every second, as his benchmark, or ten times the rate at which commercial solar cells were currently produced. It was a DARPA-hard challenge, to be sure. But when Kirkpatrick launched the project in 2005, one bold group of researchers was ready to take it on.

The group, consisting of some fifteen universities and private laboratories, has its headquarters at the University of Delaware in Newark, Delaware. There it is headed by electrical engineering professors Allen Barnett and Christiana Honsberg. Barnett is the grand old man of the university's Solar Power Program. He came to the university in 1976, intending merely to serve out the term of a three-year appointment. Instead, he stayed seventeen years. His driving passion is teaching. "It's probably my favorite thing to do in the world," he told me when I visited the university in May of 2008.

Barnett earned his Ph.D. in 1966, when he was just twenty-five years old. A few years afterward the energy crisis of the 1970s convinced him that the country's energy future lay in solar power. It's been the focus of his career ever since. When Barnett got into the field, there was no commercial market for solar cells—they had been invented in 1955 for satellites. That was still the only real market for them, and they were horribly inefficient. Barnett's initial dream was to boost solar cell efficiencies to 8 percent.

Christiana Honsberg started her career as Barnett's student. She's a cheerful but hard-driving fortysomething who grew up in the Newark area. She managed to graduate from high school at the age of sixteen. She didn't much like high school, she told me during my visit, and it "seemed sort of pointless to stay there." Both of her parents were Ph.D. chemists working at DuPont, the company now paired with the solar power team to make sure that whatever the group came up with met Kirkpatrick's requirements for manufacturability. Because she knew enough about chemistry to dislike it, Honsberg instead chose electrical engineering as her field at the University of Delaware. "From electrical engineering, you can go do just about anything else," she explained to me. She discovered she liked it and so she stayed on to earn her Ph.D., like Barnett before her, at the age of twenty-five.

Honsberg and Barnett had both spent their careers experimenting with mixing different types of materials in solar cells in an effort to increase their efficiency. This went against the grain in the field, where the byword is usually cost, not efficiency, and mixing materials drives up cost.

To generate electricity, solar cells need light-absorbing materials from which electrons are knocked loose when the light strikes them. The free-flowing electrons are then harnessed to produce direct-current electricity. Commercial solar cells are typically made of silicon, the same material as computer chips, because the material is cheap and it does the job of converting the visible portion of the sun's energy well enough. Research into an emerging breed of so-called thin-film solar cells makes use of other materials, but few researchers bother to combine different materials to increase efficiency.

That's a shame from Honsberg and Barnett's point of view, because, as Honsberg noted, "Thermodynamics tells you that you need more than one material." So what if that approach is initially more expensive? "The thing is," she said, "if you look at [only] how to get cost down, you limit a whole range of options that can lead you to high performance. Whereas if you focus on performance first, then you can be clever and . . . do all the engineering stuff to get the cost down." Barnett, too, felt so strongly that performance, i.e., efficiency, should have priority over cost when developing new solar power technologies that, as he put it, "my students are not allowed to use the word *cost* on any paper or in their thesis, because that is not what I'm teaching them." Affordability, agreed Barnett and Honsberg, would naturally follow in the economies of scale that were soon to emerge in the solar power industry.

Student and mentor had parted ways when Honsberg earned her Ph.D. and joined the faculty of the University of New South Wales. Honsberg had cut her teeth on basing her solar cell designs on gallium arsenide, a material that had been developed as part of a DARPA project. But the University of New South Wales already had an expansive silicon solar cell program, so she put gallium arsenide aside for the nearly ten years that she stayed there.

Now the two professors were teamed for the first time in eleven years at the University of Delaware, with their focus on the DARPA Solar Cells project. Barnett had grown bored, as he put it to me, and left the university as a full-time professor in 1993 to work full time at a company he founded in 1986 to make solar cells. "At one time it was the fourth largest [solar power company] in the world," he told me. "But the world was much smaller. Much, much smaller." From an initial plan to grow to thirty-five people, he found himself struggling to manage seven hundred employees. "It outgrew me," he told me later. Eventually he sold the company to GE, where it became known as GE AstroPower. The University of Delaware got wind that he was again on the job market, and made a strong play to get him back. Essentially, the administrators told him he could pursue any research he wanted. That and being able to teach put him in solar cell research heaven. Now the team of Barnett and Honsberg was back at work combining different materials in an effort to increase efficiency. DARPA was betting that their approach would revolutionize solar cell design.

At the heart of the DARPA project is a molecular beam epitaxy, or MBE, machine, where Honsberg and her graduate students grow the crystals that form the basis of the project's experimental solar cells. The machine, a shiny cylindrical vacuum chamber big enough to hold a man and bristling with arm-thick input tubes, squats in a clean room on campus surrounded by racks of high-powered computers and their operator's consoles. When I visited the project, I suited up in a lab coat, overshoes, a mask, latex gloves, and a bouffant cap to go into the clean room and get personally acquainted with the machine. Inside, I watched a young researcher at work while lab manager Kjeld ("Cal") Krag-Jensen, a Danish engineer in his sixties, explained the setup to me.

Each of the input tubes surrounding the vacuum chamber contained some substance to be incorporated into the finished wafer that would form the basis of the solar cell. After placing the wafer on what Krag-Jensen affectionately referred to as "the pizza paddle" and inserting it into a tube reserved for this purpose, a technician would activate the software controlling the machine, which would then initiate the process of heating up the various materials. Allowed to condense once more into solidity in the vacuum chamber, one molecule at a time, the materials would combine and gradually build up the finished product on the wafer. Special masks, serving the same function as masking tape and made by an outside supplier to the project's computer-designed specifications, defined the exact shape the technicians want to create. With the lights turned off, as the researcher working there during my visit preferred, and the room lightly bathed in the soft glow of the computer monitors and the green status lights of the computers in their racks, the silent MBE machine looked like a science fiction robot spider in its lair awaiting prey.

Other parts of the project's labs featured more conventional chip-fabricating facilities, including a state-of-the-art silicon-manufacturing facility, a photolithography lab, lit in yellow light that wouldn't expose photo equipment, and an assembly lab with equations scrawled on a blackboard. After taking a brief detour during which Barnett walked his fluffy and very friendly little golden Lhasa apso around the campus grounds—cutting across walkways to head for interesting bushes and trees—we went up to the roof, where two of Barnett's Ph.D. students had set up a test apparatus they had helped to build.

Only three places in the world are internationally recognized as official testing laboratories for solar cells: Germany's Fraunhofer Institute for Solar Energy Systems, the National Renewable Energy Laboratory in Colorado, and the Japanese National Institute of Advanced Industrial Science and Technology. With the field of photovoltaics ramping up into something of a worldwide boom, these three labs had a weeks-long backlog for testing that made it difficult for Barnett and Honsberg's group to track their own progress. So they had their students build their own testing setup. The apparatus sat on a metal cart, built around an off-the-shelf solar tracker. The tracker's electric motors centered a chip to be tested in the full light of the sun while the student-built test instrument measured the chip's energy-absorbing properties. Barnett seemed every bit as proud of the students as he was of the device they had created when he introduced me to them on the rooftop.

By the summer of 2007, the team at the University of Delaware had unofficially demonstrated efficiencies in the energy-gathering ability of their experimental solar cells of an astonishing 42.8 percent. The key breakthrough came with a suggestion by DARPA's Kirkpatrick: use so-called dichroic mirrors.

A material that generates the most power from particular wavelengths of solar energy is poor at producing power at other wavelengths, so the ideal solar cell would incorporate many different types of materials to provide maximum output for all wavelengths. Promising research into new ways to combine different types of materials was the genesis of Kirkpatrick's solar program. But the optimal set of materials that the University of Delaware team came up with couldn't, in fact, be combined. Formed into the crystals needed for solar cells in the team's MBE machine, each material did its job of absorbing particular wavelengths of solar energy admirably well. But the materials wouldn't form crystals together, which prevented them from simply being combined into one super energy-absorbing material.

Since the materials had to be separated, why not separate the sunlight entering the device as well? Dichroic, or wavelength-splitting, mirrors would do the job, and without drastically cutting down the amount of light striking the energy-gathering materials, as conventional prisms did. Dichroic mirrors allowed a full 93 percent of the light entering the system to reach the energy-absorbing materials, making them ideal for the purpose. In the design that the Delaware team came up with, sunlight enters the device, strikes a dichroic mirror, and splits into high-energy shorter wavelengths and less energetic long wavelengths. The shorter wavelengths strike indium gallium phosphide and gallium arsenide solar cells, while the longer wavelengths hit more traditional silicon, as well as indium gallium arsenide solar cells. All of which greatly boosts the energy-gathering potential of the system as a whole.

To ensure that the cells would be manufacturable in the quantities and at the price Kirkpatrick wanted, the solar power project brought in DuPont, based in Delaware, not far from the university. Working together, engineers at both the university and the chemical company would determine the best possible configuration of dichroic mirrors and light-absorbing materials to manufacture cheaply in bulk. If the project met DARPA's typically ambitious timetable, solar cells packing enough punch to power handheld military gizmos and lighten the loads of hundreds of thousands of soldiers would be ready for production by 2011.

But solar power was only one piece of the picture that made up Kirkpatrick's vision for the future of energy production. Electricity, from solar or any other source, wouldn't fit the bill for applications requiring what Kirkpatrick termed "high-energy density." Things such as trucks, airplanes, and ships couldn't run on electric motors, at least those of any reasonable size, weight, and utility. Their power needs were just too great. "It's kind of hard to think of a solar boat," Kirkpatrick said by way of an example. To achieve complete energy security with supplies that could be produced domestically—without relying on potentially hostile foreign nations—Kirkpatrick turned to biofuels.

Biofuels are liquid fuels made from organic feedstocks such as vegetable oil. They were such an obvious solution to the problem of providing the density of power in a domestic source that solar power couldn't deliver that Kirkpatrick couldn't actually remember coming up with the idea when I asked him about it. But he did remember pitching it to Tony Tether. Tether was a skeptic when it came to biofuels, as Kirkpatrick recalled. "This is just a bunch of people making a bunch of hooey," Kirkpatrick said, paraphrasing Tether's response.

Kirkpatrick knew this certainly wasn't true. Biofuels as a source of power for combustion engines have a history almost as long as the combustion engine itself. Rudolf Diesel, inventor of the diesel engine, experimented with peanut oil in his engines shortly after the turn of the twentieth century, at the behest of the French government, and concluded that the oil held promise for making the French colonies in Africa energy self-sufficient. "One cannot predict what part these oils will play in the Colonies in the future," Diesel wrote shortly before his death in 1913. "In any case, they make it certain that motor-power can still be produced from the heat of the sun, which is always available for agricultural purposes, even when all our natural stores of solid and liquid fuels are exhausted."

Biofuels have two distinct advantages over petroleum-based fuels. The main one, from a military point of view, is that they can be produced anywhere crops can be grown, and at a consistently high level without depleting nonrenewable supplies. But since, when burned, they release carbon dioxide that has only recently been extracted from the environment by plants or animals, rather than carbon that had been trapped in the Earth for millions of years, they contribute little—even after accounting for cultivation and manufacture—to the heat-trapping carbon dioxide in the Earth's atmosphere. In other words, biofuels are good for the planet as well as for the military.

Kirkpatrick bided his time, and the right moment came to push Tether for funding while they were riding together in a car after a successful meeting during which Kirkpatrick's work had been praised. As Kirkpatrick remembered it, Tether turned to him and said, "All right, that was really phenomenal. You've got a free pass. What do you want?"

"I want biofuels," Kirkpatrick told him.

Tether sighed. "You're just not going to let me get away from that one, are you?"

"No, sir," said Kirkpatrick. "I think we need to go down that path."

As Kirkpatrick told me later, "I think one of the real strengths that we have as an agency is the ability to be open-minded." No matter how harebrained Tether thought an idea was, he was willing to keep listening to a program manager who kept pitching it to him. Occasionally he could be convinced that maybe the idea wasn't quite so harebrained after all. "It's painful, and sometimes it takes a very long time," said Kirkpatrick, "but it does happen. And at this point in time, I think it is easily true to say that Tony is one of the strongest proponents of biofuels in the entire DOD."

Perhaps the most demanding application for petroleum-based

fuel is jet fuel, so Kirkpatrick decided to start there with his biofuel program. Jet fuel has to pack a lot of punch in as small a volume as possible, to reduce the amount of weight an aircraft has to lift, and the fuel has to remain fluid in the subzero air temperatures at high altitude. That last requirement was a particularly hard nut to crack in a biofuel.

Conventional biofuel can power unmodified diesel engines, but only in temperate climes. Biofuels have an unfortunate tendency to gel at much higher temperatures than their petroleum-based counterparts. Try to run a car on straight biodiesel in subfreezing temperatures, and your engine will quickly choke and die, if it starts at all. That's because conventional biodiesel is composed purely of long-chain fatty acids. All those long chains have a hard time flowing properly in cold weather; they have a tendency to bunch together, making a tangled mass that has difficulty moving briskly through a fuel line. Suppliers get around the problem by distributing biodiesel for use in cars and trucks as a blend with petrodiesel. The petrodiesel lends its ability to keep flowing at low temperatures.

Biodiesel blends could certainly help ease the pressure on petroleum supplies, but that wasn't good enough for Kirkpatrick. Energy security meant energy security, not simply conservation. So naturally he figured he'd first have to tackle the DARPA-hard problem of creating biofuel that remained nicely fluid at very low temperatures while delivering a high concentration of energy for its volume—the properties demanded of jet fuel. If his contractors could create bio jet fuel, they'd certainly be able to come up with fuels for less demanding applications, such as powering cars, trucks, and electric generators.

Mind you, just producing a biofuel that could power a jet engine wouldn't be enough. It had to be a fuel that could power an *unmodified* jet engine. "The most important thing in aviation fuel is that you run with the engine sets that are out there," Kirkpatrick explained to me, "and that's because of the tremendous capital investment that's already present." A bio jet fuel wouldn't be much good to the air force if it could run only in customized engines.

Hence the requirements for Kirkpatrick's BioFuels program. Step one: create military-grade jet fuel, called JP-8, from renewable feedstocks such as vegetable oil, in quantities great enough to test extensively. To meet the standard, the fuel would have to remain fluid down to minus fifty degrees Fahrenheit. One hundred liters, or about twenty-six gallons, should be enough to test it. Step two: demonstrate how to create it in large enough quantities and cheaply enough to make a significant contribution to the armed services' energy needs. Anything less than three dollars a gallon in quantity would make Kirkpatrick and Tether happy. Not incidentally, JP-8 powered everything from generators to Humvees as well as jets, further increasing its utility to the military. It went without saying that any commercial supply of biofuel produced as a result of the BioFuels program would also make a significant contribution to the nation's broader energy needs.

Kirkpatrick chose three performers to work on the BioFuels program. "That's a classic portfolio-based approach to technology development," he explained to me, and it was one DARPA used frequently. As Kirkpatrick put it, the agency funds "complementary technical approaches to reduce risk."

The players in this case were the Energy and Environmental Research Center, or EERC, based at the University of North Dakota, in Grand Forks, North Dakota; General Electric; and an Illinois-based division of Honeywell called UOP. EERC's biofuel underwent testing at Wright-Patterson Air Force Base in January 2008, and passed with flying colors. After that, it was on to step two for them: demonstrate the technologies for commercial-scale production.

UOP, too, had produced their fuel in great enough quantities to test by the time I sat down to write this book, but EERC led the pack because it had already been working on high-grade biofuels on its own and for the air force for fully fifteen years before the BioFuels program got under way. All DARPA had to do was give an additional boost to work that was already in progress. "DARPA stepped in with a large enough amount of money to really close this quickly," Kirkpatrick told me. And he said something else, something that I'd never heard a DARPA program manager say of any DARPA contractor: "They have perhaps some of the most innovative technology I've ever seen."

Of course, then I had to see for myself what was going on at an obscure research center in an out-of-the-way corner of the North Dakota plains.

"They're a very dry sort of North Dakota folks," Kirkpatrick warned me before my trip. "Don't be surprised if you get a lot of understatement."

THE EERC PR REP warned me that the five-hour drive from Minneapolis to Grand Forks was brutal, but I didn't see what he meant until I was on the road for two or three hours in April 2008. It was just the same flat plains land for mile after mile after mile, dotted at intervals with farmhouses and barns and the occasional one-gas-station town to break up the monotony.

The University of North Dakota seemed to take up most of the town of Grand Forks. I found my way easily to the low modern buildings of the campus, but lost my way in the sameness of it all as I hunted for the Energy and Environmental Research Center.

An American flag whipped in the wind on its pole outside the glass-and-stone facade. There was no one else around as I went in. The PR rep was back in Minneapolis for the annual BioFuels Conference, so he wasn't there to greet me. Instead I met research manager Tom Erickson, his deputy, Chris Zygarlicke, and Gerald Groenewold, the director and founder of the EERC, in Groenewold's office.

The three were indeed understated in their manner, but there was no masking the fervor with which Groenewold and his colleagues attacked their work. They were on a mission to rescue the world from the environmental destruction wrought by the unchecked consumption of fossil fuels and to develop commercially viable alternatives.

Commercial viability was central to the EERC's mission. Groenewold didn't think much of theoretical fixes that did no good in the real world outside of a secluded laboratory. Although EERC was housed at a state university, he ran the place like a corporation. It is not, he likes to emphasize, an academic institution. Neither he nor the people working for him have any teaching duties. Instead, they are entirely focused on applied research geared toward practical applications. "We're state employees with no state money by choice," he told me, "and that makes us extremely entrepreneurial." In fact, fully 80 percent of the EERC's funding comes from private companies that hire the group to develop specific, practical technologies. "This place is very Darwinian," Groenewold said. "Those who can, do. Those who can't, leave. If you don't bring in money, you don't get a paycheck. It's as simple as that."

To hear Groenewold tell it, the university simply provides land and buildings, and keeps the hell out of the EERC's way. "My definition of peer review is very different from the other side of campus," he told me. "It is somebody in the private sector investing their cash in what we're doing. If they're not investing their cash in what we're doing, why are we doing it? That's peer review here."

Groenewold's office was expansive, and included a conference table. He worked at a wide desk beneath a wall of windows and a Native American dream catcher. A framed picture on one wall reproduced an antique photo of Native Americans holding rifles, with the caption "Homeland Security: Fighting Terrorism Since 1492." A glass-topped coffee table in the conference area held innumerable fossils that Groenewold had collected over the years.

Gerry Groenewold was on the late side of middle age, lean, with glasses, white hair, and a matching goatee. During our meeting he wore a checkered sport jacket, a white dress shirt with no tie, and slacks with a big silver belt buckle. Although soft-spoken, he talked forcefully and passionately on the subjects closest to his heart: energy and, as he saw it, the inextricably connected natural environment.

Groenewold launched what became the EERC in 1974, only a couple of years out of engineering school with a Ph.D. from the University of North Dakota (he is a hydrogeologist and geochemist by training), with a grant from the Environmental Protection Agency. "I had like five months of salary, and a dream," he told me. He was then twenty-eight years old.

He had remained at the university after graduating, not as a faculty member, but as an employee of the State Geological Survey, which had a lab there. But he grew dissatisfied with life as a conventional government employee, and wanted to push for something more. Something more entrepreneurial. When the Arab oil embargo threatened to bring the Western economies to their knees, the time seemed right for a radical move for a hydrogeologist with an abiding concern for the environment. Groenewold was then only one of countless engineers, scientists, and entrepreneurs who hoped to develop renewable energy sources and clean up the environment at the same time. It wasn't too difficult for him to persuade the university's dean of engineering to let him set up shop as head of a small, semiautonomous organization. "I just started bringing in money from contracts," and he consistently turned down state funding that wasn't awarded on a competitive basis, he told me. He has upheld the practice ever since.

After taking over a federal research-and-development center based at the university in the 1980s, he transformed it into the $227-million-a-year entrepreneurial research center that he now leads. Today, says Groenewold, the time is ripe, much more so than in the early 1970s, for a revolution in green power. "Irrational exuberance—we had a lot of that going on then," he says of his early days in the field. "The bottom line was . . . well, the bottom line was not considered. That's the bottom line."

Things are different this time around. For instance, Groenewold and his colleagues know that fossil fuels aren't going to go away anytime soon. That's why one of his group's major projects is to help develop the technologies needed to build clean-coal power plants. Ever ambitious, he now sees the EERC as well positioned to become one of the top organizations in the world developing and building and retrofitting coal-fired power plants that will realease no harmful emissions, for instance carbon dioxide.

Staunching the flow of carbon into the atmosphere is of utmost importance to Groenewold and the EERC, as is protecting the world's clean water supply. "Water is not a resource," Groenewold asserted to me. "It is *the* resource." With an ever-increasing demand for the world's water supplies—including for power plants, which Groenewold cites as the biggest user of water after agriculture—the issue of water is going to be increasingly important in the twenty-first century, along with myriad other environmental concerns.

"I'm an environmentalist, but I'm an environmentalist with my feet on the ground," said Groenewold. "Energy and environment are inseparable. You don't do one without considering the other." At the time of our conversation, the EERC was working with $315 million in funding from various corporate partners on technologies for carbon sequestration—that is, extracting and storing atmospheric carbon dioxide—in a ten-year program. The group was also big on removing mercury from all industrial emissions, not just coal-fired power plants. "Nobody does more work in mercury," Groenewold told me.

Groenewold continues to make sure no state money comes in the door that the team doesn't have to compete for with other institutions or companies. Why? "Freedom," he told me simply. Freedom to take the kinds of risks that Groenewold feels wouldn't be acceptable in a government-controlled organization. Taking noncompetitive government handouts, Groenewold has always felt, would put him on a slippery slope. "That's the beginning of micromanagement," he said.

On the subject of working at the EERC—"You have to be creative and have a fire in your tummy," says Groenewold, "and want to do something useful with your life instead of sit around and suck up oxygen all day." Harsh as that might sound, Groenewold in fact treats the people working under him very well. "A little rule here," he said. "If you have kids and your kid has a soccer game, the only reason I'm going to be upset is if you don't go."

Groenewold, who does a lot of his recruiting from the university, takes a rather unorthodox approach to hiring. He asks each potential hire who makes it through the initial screening process to answer three questions:

"What is your dream?" That's to sort those with fire in their tummy from the oxygen suckers, as Groenewold puts it.

"Do you feel lucky?" Groenewold wants the people who work at the EERC to create their own luck and opportunities rather than wait around for someone to do it for them.

And, finally, "How many speeding tickets have you had?" That's to filter out people who are overly risk averse. "I think a person who has never had a speeding ticket is so risk averse," Groenewold explained to me, "that they should not work here."

"I don't think we'll ever be energy independent," Groenewold said, "but I think we can, and we must, focus on being energy secure. There is a big difference. To say we want to be energy independent to me sounds like an insult to our Canadian partners and other partners and people we really trust and value as partners." Groenewold also agrees with Kirkpatrick's assessment that there is no single solution to energy security. The solutions, he feels, will come in the form of a range of technologies. In addition to biofuels made from vegetable oil and algae and carbon sequestration technologies, the group is also developing small, portable power plants suitable for powering factories or small communities with any available biomass. Simply throw agricultural waste, wood chips, sawdust, grass, or anything else available inside, and let the system cook it slowly, producing gas that could then run a generator. Flexibility in the type of feedstock to be used is key, so that these so-called distributed (as opposed to centralized) systems can adapt to the particular geography or feedstock they find at hand.

The EERC is also looking at hydrogen for powering automobiles. Of course, hydrogen fuel is on a lot of people's minds these days, but the EERC has a novel approach to solving the problem of producing hydrogen without contributing to the greenhouse effect. At the same time it has solved the problem of distributing the hydrogen. True to EERC's general trend away from the centralized distribution of power, the group's engineers have proposed producing hydrogen right at the fuel pumps. A small system would produce hydrogen from water through electrolysis driven by wind energy on demand, when a driver squeezed the pump handle to fill up his car.

After our talk, research manager Erickson gave me a walking tour of the EERC. In the main building I saw labs full of glass beakers and tubing, and a scanning electron microscope for examining the buildup of particulates on the inside of coal-fired power plant smokestacks. In other buildings I put on a hard hat and watched technicians at work on experimental power plants, including one that ran on sewage. Presumably, the EERC's emission-scrubbing technology would see to it that smells as well pollution would stay out of the atmosphere when that one was running. Everything was enclosed and protected from the harsh North Dakota winters, except for a portable power plant, in a trailer, that ran on wood scraps. It was a working demonstration of the EERC's plan for such generators, and they could pull the thing up to a construction site or factory making wood products, plug it into whatever equipment needed power (the office trailers, for example), and run the whole operation on the scraps produced at the site. The one thing Erickson wouldn't let me get a good look at was the setup for processing straight vegetable oil into jet fuel.

At the time of my visit, the EERC was engaged in the second part of DARPA's BioFuels assignment: demonstrate that it could produce its JP-8 biofuel in commercially viable quantities. The group had succeeded in the first stage of the project, producing the fuel itself, but the exact method by which the EERC produced its JP-8 from vegetable oil was a closely guarded secret. Not a military secret, but a proprietary one. Groenewold didn't want any upstart companies or organizations freeloading off his group's years of hard work. He did tell me that the EERC's method more closely resembled conventional oil refining than the standard chemical processes used in biodiesel manufacturing.

"We approached it more from a refining process," Erickson reiterated.

Refining the feedstock through thermal processing allows the engineers to lower the finished product's gelling temperature without further modification. "We don't have any additives to get lower temperatures in our fuel at all," said Erickson. The basic process in the EERC labs involves piping vegetable oil through a so-called reactor about the size of a refrigerator. After an hour, out comes five liters of what Groenewold and his colleagues term "green diesel." They don't like the term *biodiesel* because of its association with the well-known limitations of that type of fuel.

At the chemical level, the EERC process takes those problematic long-chain fatty acids that cause gelling at high temperatures and isomerizes them. That is, it rearranges their molecular structures, in this case into a more or less spherical configuration resembling snowflakes with their myriad arms. Those nicely organized shapes can much more easily flow through a vehicle or generator's fuel delivery system and into an engine. That isomerization of the triglyceride—that is, vegetable oil or other fatty acid—on the way toward creating the biofuel is the crucial step needed to turn mere biodiesel into military-grade JP-8.

"It is chemically indistinguishable from what you would get from petroleum," DARPA's Kirkpatrick told me of the bio jet fuel both EERC and UOP had developed. "It's JP-eight." In fact, asserted Kirkpatrick, the new fuel—more than half of which was made up of isomerized molecules—was *better* than petroleum-based JP-8, which has only a 50 percent concentration of isomerized molecules.

Even so, Kirkpatrick and his colleagues went to ASTM International, the standards organization formerly known as the American Society for Testing and Materials, to certify their JP-8, and ran into a wall. The organization wanted to classify DARPA's JP-8 as a biofuel, not as straight-up aviation fuel. "We're not a biofuel," Kirkpatrick told ASTM officials. "We're JP-eight from bio feedstocks." As proof, Kirkpatrick pointed to a gas chromatograph/mass spectrum representing all the molecules in the new fuel. "Every single molecule we have is in your JP-eight," Kirkpatrick told the ASTM, "and we satisfy all your other specs" for JP-8.

"But you come from soy oil," an ATSM official complained.

"Yes. What's your point? You're a scientist. You understand what a gas chromatograph/mass spec looks like, right? There's the molecule. We have less bad things in our fuel than you have in petroleum."

"Yes, but you're from soy oil."

And so it went.

Now that the challenge of producing the stuff in the lab had been overcome, DARPA faced the perhaps even greater challenge not only of producing the stuff in quantity, but of persuading the operators of all those airplanes and trucks to use it. "I think when we fly it—when we actually can convince somebody to fly one hundred percent biofuel and somebody to agree that we've done one hundred percent bio jet—and somebody signs up to actually produce it at three dollars a gallon or less," Kirkpatrick said, "then I think it will be big news. I'm okay that it's not big news yet." After all, only half of Kirkpatrick's goals had been met. "It's almost soup," he said. "If you're in the kitchen, it smells damn good. But I also know it's not done yet. It's got a ways to go."

One advantage to the vegetable jet fuel is that it doesn't matter what kind of vegetable oil the researchers start with. Rapeseed (i.e., canola) oil, safflower oil, animal renderings, or even algae can be used equally effectively. "If it's a triglyceride," said Kirkpatrick, "we can turn it into jet fuel." Most biofuels, including those made by the EERC, are still made from vegetable oil, which comes from crops such as soybeans and corn. Put enough of that grain production into fuels instead of food, and you get problems. In the spring of 2008, the time of my visit to the EERC in North Dakota, worldwide food shortages were being blamed in part on increased biofuel production in Europe and the United States. As Gerry Groenewold told me during my visit, "We're going to have to move away from using food to produce energy."

Next-generation biofuels—fuels made from nonfood sources that could be developed without taking over land that would ordinarily be devoted to food crops—were around the corner, but hadn't yet broken through. Kirkpatrick aimed to speed that process with new programs to develop fuels from so-called cellulosic feedstocks—that is, feedstocks composed of inedible sources such as corn husks, grasses, and algae. By Kirkpatrick's calculation, cellulosic feedstocks alone could feed fully half of the nation's appetite for transportation fuels. This includes things such as paper waste and other discarded materials. The big challenge here is getting energy from these tough-to-break-down sources, and doing it affordably enough on a large scale.

Beyond cellulosic biofuels, Kirkpatrick planned to foster development of commercial-scale algae-based fuel. EERC is ahead of the curve there, too, already coming up with ways to use its carbon dioxide–capture technology to pipe carbon from coal-fired power plant emissions through algae farms that would need only carbon dioxide and sunlight to grow.

Beyond biofuels in all their forms and solar power, Kirkpatrick also wanted to develop portable wind power in the form of generators no larger than a bread box that could be carried into the field by special ops forces, as well as miniature hydroelectric generators that could generate electricity from mountain streams, and even microgeothermal generators that could produce power by capitalizing on the difference between aboveground and underground temperatures. The last system, if it could be developed, would most likely generate only minuscule amounts of power, but it could be ideal for powering sensors that would sit unattended—and it is hoped, undetected—for long periods of time in a semidormant state, waking up only long enough to squirt out a quick radio transmission when it detects something interesting.

Ultimately, Kirkpatrick wanted to develop the means by which any size group of military forces—from the single soldier to entire battalions and installations as large as air bases—can become entirely energy self-sufficient. Solar panels and fuel cells could fit the bill for individual soldiers, while filing cabinet–size biofuel generators could supply small encampments, possibly operating from the back of a Humvee. To be most effective, the generators would have to be able to run on whatever biomass was available. In his DARPATech presentation, Kirkpatrick pointed to the 1985 movie *Back to the Future*, in which the heroes power their time machine with ordinary banana peels and other garbage, as a model for how these systems might work. "If bananas aren't in season," Kirkpatrick quipped, "they should be able to use other cellulosic material."

For large installations, Kirkpatrick envisioned using algae bioreactors, in which algae would grow in transparent containers, using sunlight and carbon dioxide to multiply, and ultimately generate power. Such a system could sustain itself indefinitely. A more or less permanent base could even construct an open algae pond to produce six to eight times as much oil-rich algae as a portable system.

The idea is to have a constellation of power systems at a commander's disposal, to take advantage of any kind of weather conditions or terrain. "Sometimes the sun won't shine," said Kirkpatrick. "Sometimes there's a drought and crops don't grow. We need equipment that will work anywhere, anytime. If we go in with batteries, maybe we take two solar rechargeable batteries and one fuel cell—whatever works."

Kirkpatrick himself is betting his career on the renewable energy technologies he fostered at DARPA. With a year left on his badge for working at the agency, in March 2008 he handed over his Very High Efficiency Solar Cells to another program manager. "I have to be separate from any matter that I might have taken interest in as a [DARPA] manager for a year before I can go back out to that industry," he explained to me of the government regulations that required him to avoid any conflict of interest. And he sees the emerging solar power industry as the best place to be working in the near future. In fact, Kirkpatrick believes that solar power will eventually account for 75 percent of all power generation in the United States.

And he doesn't mean just power feeding fixed installations such as houses, but that being used for transportation as well. It might take a while—within twenty to thirty years is Kirkpatrick's guess—but it will happen. It has to happen, given the ever-rising cost of producing and securing petroleum. Already, Kirkpatrick told me, with commercially available, comparatively inefficient solar cells, it is possible to build electric cars that run for less than petroleum-powered vehicles burning gasoline that costs three dollars per gallon. "We are already under three dollars per gallon," Kirkpatrick said. "But what is surprising me is that nobody has figured that out yet."

In Kirkpatrick's vision of the future, the southern and western states would export energy to the northern states. Even so, the northern states will still be able to generate significant amounts of their own power. Kirkpatrick looks to Germany as an example of how to make solar power viable on a national scale; even that northern nation sees enough sunlight to make solar power a significant contribution to the national grid.

The German government gives tax breaks for installing solar power where it is needed and saves money nationally by not installing additional transmission lines that rising electric use would require if that additional power were generated by conventional coal plants and other centralized power generation systems. "High-powered transmission lines cost a million dollars a mile," Kirkpatrick pointed out. "So they spent something like eight billion euros subsidizing solar cells going in, but they saved twelve billion euros in not building transmission line capacity. We're a little bit behind the times, but we'll catch up."

And that was with commercially available solar cells available in the late 2000s, before Kirkpatrick's solar cells were on the market. I asked him how his cells would change the equation. His feeling was that the fully 50 percent–efficient cells his program had set out to create would still be too expensive for ordinary use, at least at first, so they would go to mostly military uses. But advances made through the program would result in consumer-grade solar cells twice as efficient as today's commercial cells, for the same price. Those 35 percent–efficient cells, once installed on rooftops, would yield electricity for one dollar a watt, or three to five cents per kilowatt hour—a cost comparable to hydropower. But that hydropower equivalent could be tapped wherever it was needed, not just where a river or a waterfall happened to be or could be created.

In Kirkpatrick's view, such distributed power generation, not the mostly centralized systems we have today, would carry the nation through the next century. If he is right, we're right around the corner from a revolution in power generation analogous to the switch that occurred starting in the late 1970s from mostly mainframe computing to the personal computers now in individual homes and businesses in the millions. Big power plants would still be needed, solar and otherwise, to transmit electric power to intemperate climes that didn't see as much sun as other places, but they would be only one part of the energy equation, not most of it.

ADVANCES IN POWER generation may well join past DARPA projects such as the ARPANET in forever changing the way we live and work. But we must not take these and other marvels for granted. They have to be conceived and developed by dedicated managers, scientists, engineers, and other visionaries. Visionaries such as those working for the Defense Advanced Research Projects Agency.

One program manager described DARPA to me as a national treasure, and I tend to agree. However, I believe it will stay that way only if, along with giving the agency its due, we also practice the benign neglect that it requires to thrive.

DARPA was conceived and has operated, first and foremost, as an agent of change for the United States armed services, and I believe it should stay that way. Its mission to equip our nation's war fighters with a technological edge over their adversaries gives DARPA a razor-sharp focus it would otherwise lack, and its emphasis on quickly moving projects from concept to working prototype while making as efficient use of funds as possible should be a model for research and development throughout the Department of Defense.

Even more than that, DARPA should also be a model for research and development throughout the federal government. It has proven again and again that true innovation need not depend on massive expenditures and armies of bureaucrats. Operating on one half of 1 percent of the U.S. defense budget, with a staff housed in a single office building of modest size, DARPA has fostered some of the most useful technological innovations of all time. Key to its success—in addition to its minimal bureaucracy—has been the term limits for its managers and its low overhead. The term limits ensure that the people who do its most important work care more about fulfilling the agency's mission than protecting their jobs. The decision made at the outset that the agency shouldn't maintain its own laboratories but instead farm out work to other organizations has allowed it to consistently stay ahead of the technological curve by quickly developing new capabilities and letting others go as the need arises. Managers of any new government agency seeking to capitalize on DARPA's recipe for success—for instance, the Advanced Research Projects Agency–Energy, or ARPA-E, funded under the Obama administration in 2009—would do well to remember these essential ingredients.

Here's to the next fifty years.

NOTES

INTRODUCTION

xiii   The autodoc: Larry Niven, *World of Ptavvs: Three Books of Known Space*, New York: Ballantine Books, 1966, 1968, 1975, 1990, renewed 1994, 1996, p. 150.

xiii   forty-seven-year-old mechanical engineer: Thomas Low, Director, Medical Systems and Devices Department, Engineering and Systems Division at SRI International. See chapter 4.

xv   Retail price of a Boeing 737-600 airliner: "Commercial Airplanes: Jet Prices," Boeing Web site, http://www.boeing.com/commercial/prices/index.html (accessed October 16, 2008).

xv   *SpaceShipOne* and commercial space flight, including XCOR and AirLaunch LLC: See Michael Belfiore, *Rocketeers: How a Visionary Band of Business Leaders, Engineers and Pilots Is Boldly Privatizing Space*, New York: Smithsonian Books/HarperCollins, 2007.

xv   My visit to XCOR and related quotes: I visited XCOR on June 22, 2004, and wrote about it in my story "Commercial Spaceflight Takes Off in Mojave," *Chronogram*, October 2004, online at http://archive.chronogram.com/issue/2004/10/feature/ (accessed March 24, 2009).

xvii   XCOR's tea cart rocket: "XR-2P1, 15lbf N2O/Ethane Tea Cart Engine," Web site of XCOR Aerospace, http://www.xcor.com/products/engines/2P1_N2O_ethane_rocket_engine.html (accessed May 17, 2008).

xvii   Rocket company with $30 million in financing: Gary Hudson, former Rotary Rocket CEO, e-mail correspondence, April 17, 2009.

xviii   XCOR's piston pump: See "XCOR Rocket Propellant Piston Pumps," Web site of XCOR Aerospace, http://www.xcor.com/products/pumps/index.html (accessed March 5, 2008); "XCOR Completes DARPA Contract," XCOR Web site, http://www.xcor.com/press-releases/2004/04-10-21_XCOR_completes_DARPA_piston_pump.html (accessed March 5, 2008);

and http://www.xcor.com/press-releases/2003/03-09-26_XCOR_DARPA_ pump_milestone.html (accessed March 5, 2008).

xxi   DARPA budget: "RDT&E Programs (R-1)," Department of Defense Budget, Fiscal Year 2008, U.S. Department of Defense, DefenseLINK, http://www .defenselink.mil/comptroller/defbudget/fy2008/fy2008_r1.pdf (accessed March 24, 2009), p. D-2.

xxi   U.S. defense budget: "The Budget for Fiscal Year 2009," U.S. Government Printing Office, GPO Access, http://www.gpoaccess.gov/usbudget/fy09/pdf /budget/defense.pdf (accessed March 24, 2009), p. 49.

xxi   U.S. defense spending is more than half of the U.S. discretionary budget: "The FY 2009 Pentagon Spending Request—Discretionary," Web site of the Center for Arms Control and Non-Proliferation, http://www.armscontrolcenter .org/policy/securityspending/articles/fy09_dod_request_discretionary/ (accessed March 24, 2009).

xxi   U.S. defense budget more than that of China, Russia, and Europe combined: "SIPRI Yearbook 2008: Armaments, Disarmament and International Security—Summary," Stockholm International Peace Research Institute, online at http://yearbook2008.sipri.org/files/SIPRIYB08summary.pdf (accessed March 24, 2009), pp. 10–11.

xxi–xxii Eisenhower speeches: "Dwight D. Eisenhower Speeches," Web site of the Eisenhower Presidential Library and Museum, http://www.eisenhower.archives .gov/All_About_Ike/Speeches/Speeches.html (accessed March 22, 2008).

xxiii   B-2 bomber cost: "B-2 Bomber: Cost and Operational Issues," United States General Accounting Office report reproduced on the Web site of the Federation of American Scientists, http://www.fas.org/man/gao/nsiad97181.htm (accessed October 16, 2008).

xxiii   NASA budget: "NASA Unveils $17.6 Billion Budget," Web site of the National Aeronautics and Space Administration press release, online at http://www .nasa.gov/home/hqnews/2008/feb/HQ_08034_FY2009_budget.html (accessed October 16, 2008).

01: AN ARM AND A LEG

1   Kuniholm's firefight and recovery: "John [sic] Kuniholm's an Ordinary Man with an Extraordinary Story," *Assistive Technology News*, http://www .atechnews.com/images/John_Kuniholm.pdf (accessed March 24, 2009); "Account of a Firefight," e-mail by U.S. Marine major Thad Coakley reprinted by Donald Sensing in his One Hand Clapping blog, http://www .donaldsensing.com/?p=424 (accessed December 10, 2008); Jonathan Kuniholm, "Open Arms," *IEEE Spectrum*. March 2009, p. 37, and e-mail correspondence, May 29, 2009.

2   Myoelectric versus body-powered arms: YouTube user theguywiththehook, "Prosthetic Options: Myoelectric Hook and Hand," http://www.youtube .com/watch?v=bCSVkkhSy3M (accessed March 24, 2008).

3   Myoelectric arms: Michael Weisskopf, *Blood Brothers*, New York: Henry Holt and Company, 2006, pp. 125–6.

3   Kuniholm's background and his prosthetics projects: Sam Boykin, "With Open-Source Arms," *Scientific American*, October 2008, p. 90.

4   Geoffrey Ling quotes and background: Geoffrey Ling, telephone interview, March 26, 2008.

8   Kuniholm's 2008 Democratic National Convention speech: Jon Kuniholm, "2008 Democratic National Convention: Remarks as Prepared for Delivery by Jon Kuniholm, Wounded Iraq Veteran," online at http://www .prnewswire.com/cgi-bin/stories.pl?ACCT=104&STORY=/www/story/08-28-2008/0004875460&EDATE= (accessed December 11, 2008).

8   Oscar Pistorius: Josh McHugh, "Blade Runner," *Wired*, March 2007, online at http://www.wired.com/wired/archive/15.03/blade.html?pg=1&topic=blade &topic_set= (accessed March 24, 2009).

10   100,000 of 1.7 million American amputees are missing part of an arm: Sam Boykin, "With Open-Source Arms," *Scientific American*, October 2008, p. 90.

10   "just not that interested": Agnes A. Curran quoted by Sam Boykin in "With Open-Source Arms," Ibid.

10   "Prosthetics is one of many underserved markets in which innovation has stagnated," Jon Kuniholm, quoted by Sam Boykin in "With Open-Source Arms," Ibid.

10   Military's interest in better prosthesis increased in 2003: Michael Weisskopf, *Blood Brothers*, New York: Henry Holt and Company, 2006, p. 122.

11–12 Jon Kuniholm on hooks, APL: I visited the APL and conducted interviews with Kuniholm and others there on August 1, 2007.

14   APL history: "About APL," APL Web site, http://www.jhuapl.edu/aboutapl/ (accessed March 24, 2009); Applied Physics Laboratory, *The First Forty Years: A Pictorial Account of the Johns Hopkins Applied Physics Laboratory Since Its Founding in 1942*, first published in book form by Schneidereith and Sons, Baltimore, Maryland, 1983, available online at http://www.jhuapl.edu/aboutapl/ heritage/firstforty/ (accessed March 24, 2009), pp. 1–5 (formation of APL for proximity fuse effort in 1942); p. 109 (work on Transit).

15     $100 million project: Jon Kuniholm, "Open Arms," *IEEE Spectrum*.

15   Kuniholm's involvement with Revolutionizing Prosthetics: "About," Web site of Tackle Design, http://www.tackledesign.com/about.php (accessed March 24, 2009).

16   Stuart Harshbarger background and quotes: Stuart Harshbarger, telephone interview, April 26, 2007.

22   DARPATech: I attended DARPATech and met Dean Kamen, Chuck Hildreth, and other members of the Revolutionizing Prosthetics teams on August 7–9, 2007.

26   "Without a market, none of this means anything": Jon Kuniholm, "Open Arms," prepublication version of the *IEEE Spectrum* article.

26   "I think we can generate far more societal benefit if we give away information than if we commercialized and sold the ideas.": Jon Kuniholm, quoted by Sam Boykin in "With Open-Source Arms," *Scientific American*, October 2008, p. 90.

26   Half of DARPA's work is classified: According to DARPA public relations manager Jan Walker, the agency doesn't have the ratio of classified to non-classified programs readily available, but program manager Douglas Kirkpatrick told me in a phone interview on April 4, 2004, "You're getting an impression of probably about fifty percent of the organization," when I asked him how much of DARPA's work I was privy to.

02: A SPECIAL-PROJECTS AGENCY

29   Memoirs of Major General J. B. Medaris: *Countdown for Decision*, New York: G. P. Putnam Sons, 1960. Much of the information in this chapter comes from this book. Another primary source was Gordon L. Harris, *Selling Uncle Sam: A Firsthand Look at the Public Relations Behind the U.S. Space Program,* Hicksville, N.Y.: Exposition Press, 1976. Secondary sources included Matthew Brzezinski, *Red Moon Rising: Sputnik and the Hidden Rivalries That Ignited the Space Race,* New York: Times Books/Henry Holt and Company, 2007; Michael J. Neufeld, *Von Braun: Dreamer of Space, Engineer of War,* New York: Alfred A. Knopf, 2007; and Katie Hafner and Matthew Lyon, *Where Wizards Stay Up Late,* New York: Simon & Schuster, 1996.

32   Vanguard "a very useful tool for space research" and related quotes and details of early launches: Wernher von Braun, "The Story Behind the Explorers," *This Week*, April 13, 1958, p. 8.

35   Lieutenant William Magill: Harris, *Selling Uncle Sam*, p. 81.

35   "General, . . . it has just been announced over the radio that the Russians have put up a successful satellite!": The quote comes from Medaris, who attributes it to Gordon Harris. Harris, however, writes in his memoirs, *Selling Uncle Sam*, that although he was the first on the base to get the news (from a newspaper reporter calling him for comment), he wasn't present at the officers' club. Instead, he reports that he phoned the news to the club, and that Lieutenant William Magill took it to Medaris, von Braun, and the others.

35   Victrola needle: Medaris, *Countdown for Decision*.

36   Von Braun's later recounting of the episode: von Braun, "The Story Behind the Explorers."

36   "prolonged nightmare": Charles J. V. Murphy, "The White House Since Sputnik," *Fortune*, January 1958, p. 98. The advisor is not named.

37   "I was particularly annoyed" and other quotes and details from Eisenhower's memoirs: Dwight D. Eisenhower, *The White House Years: Waging Peace, 1956–1961*, Garden City, N.Y.: Doubleday and Company, 1965.

39   Eisenhower press conference: "Transcript of the President's News Conference on Foreign and Domestic Matters," *New York Times*, October 10, 1957, p. 14.

41   McElroy entering his office and related details: "The Organization Man," *Time*, January 13, 1958, available online at http://www.time.com/time/magazine/article/0,9171,862823,00.html (retrieved May 1, 2008).

43   Science Advisory Committee: James R. Killian, Jr., *Sputnik, Scientists and Eisenhower*, Cambridge, Mass.: MIT Press, 1977, introduction, first page.

43   "When I was growing up, all the boys wanted to play first base. Now most of them seem content to sit in the bleachers": Quoted by Murphy, *Fortune*, January 1958, p. 98.

44   "hair-raising chemicals": "The Young Rocketeers," *Time*, January 13, 1958, available online at http://www.time.com/time/magazine/article/0,9171,862854,00.html (retrieved May 1, 2008).

44   The scientists went to McElroy: Murphy, *Fortune*, January 1958, p. 98.

45   "a lively sense of urgency": Quoted ibid.

45   "Does the Secretary of Defense have legal authority" and other quotes from documents related to the formation of ARPA: Richard J. Barber Associates, *The Advanced Research Projects Agency 1958–1974*, study commissioned by ARPA, published 1975, available online from the Defense Technical Information Center at http://handle.dtic.mil/100.2/ADA154363 (accessed March 24, 2009).

47   "looking drawn . . . nervous and grumpy": Murphy, *Fortune*, January 1958, p. 98.

49   the name under which it would start life: A lieutenant colonel, George Brown of the air force, working in a Department of Defense office, gets credit in *The Advanced Research Projects Agency 1958–1974* for coming up with the name in a footnote on page II-10.

49   one too many "Special" organizations: *The Advanced Research Projects Agency 1958–1974*, p. II-10.

53   "A temporary expediency": *The Advanced Research Projects Agency 1958–1974*, p. II-39.

53   "I don't want to be rushed into it": "The Organization Man." *Time*, January 13, 1958, available online at http://www.time.com/time/magazine/article/0,9171,862823,00.html (accessed May 1, 2008).

53   Roy Johnson and quotes: Jack Raymond, "Pentagon Forms a Space Agency;

G.E. Aid Is Chief," *New York Times*, February 8, 1958, p. 1; "Space Architect: Roy William Johnson," *New York Times*, February 8, 1958, p. 8.

54   "lasting peace here on earth": *The Advanced Research Projects Agency 1958–1974*, p. II-39.

54   Details of ARPA formation and structure: *The Advanced Research Projects Agency 1958–1974*.

56   "it should not have to justify this activity to this civilian agency": *The Advanced Research Projects Agency 1958–1974*, p. III-27.

57   Saturn rocket origins: *The Advanced Research Projects Agency 1958–1974*.

57   Transit navigation system: Catherine Alexandrow, "The Story of GPS," and Stew Magnuson, "Connection to the Cosmos," *DARPA: 50 Years of Bridging the Gap*, U.S. Department of Defense and Faircount Media Group, Tampa, Fla., 2008.

58   DDR&E and other details of ARPA's growing pains: *The Advanced Research Projects Agency 1958–1974*.

58   The resulting paper explored several options for the future of ARPA: *The Advanced Research Projects Agency 1958–1974*, p. III-71.

58   "dead cat hanging in the fruit closet": *The Advanced Research Projects Agency 1958–1974*, p. VI-1.

59   "There is just less argument with ARPA": *The Advanced Research Projects Agency 1958–1974*, p. III-63.

59   Materials science and other early ARPA research: *DARPA: 50 Years of Bridging the Gap*.

60   "we wired the seismic world together" and related quotes: J. R. Wilson, "Detecting the Threat," *DARPA: 50 Years of Bridging the Gap*, p. 58.

03: THE INTERGALACTIC COMPUTER NETWORK

63   AN/FSQ-32XD1A, its role at ARPA, and related quotes: *The Advanced Research Projects Agency 1958–1974*.

63   SAGE computer appearance and details: Bob Yelavich, "SAGE Computer (circa 1957)," http://www.yelavich.com/mphotos/sage.htm (accessed March 24, 2009); "SAGE Computer System," *History of Computing*, Lexicon Services, 1982–2001, online at http://www.computermuseum.li/Testpage/IBM-SAGE-computer.htm (accessed March 24, 2009).

64   "provide a better understanding": *The Advanced Research Projects Agency 1958–1974*, p. V-49.

64   Ruina had more pressing projects: *The Advanced Research Projects Agency 1958–1974*, p. V-7.

64   "Lick": Hafner and Lyon, *Where Wizards Stay Up Late*, 1996, p. 29.

65   Licklider's "Man-Machine Symbiosis": J. C. R. Licklider, "Man-Machine Symbiosis," *IRE Transactions on Human Factors in Electronics*, vol. HFE-1, March 1960, pp. 4–11, online at http://groups.csail.mit.edu/medg/people/psz/Licklider.html (accessed March 24, 2009).

66   "something that was important"; "best academic computer centers"; and the Intergalactic Network: *The Advanced Research Projects Agency 1958–1974*, V-51–52.

67   Engelbart, his ideas, and his ARPA-funded work: Donald L. Nielson, *A Heritage of Innovation: SRI's First Half Century*, SRI International, Menlo Park, Calif., 2004, 2006, pp. II-1–42.

67   "faster than a typewriter can achieve": Ibid., p. II-15.

68   "impossible to think of anything important": Quoted in *Advanced Research Projects Agency 1958–1974*, p. VII-34.

68   "This new way of using computers": *Advanced Research Projects Agency 1958–1974*, p. VII-33.

68   "In my office in the Pentagon": Keenan Mayo and Peter Newcomb, "How the Web Was Won," *Vanity Fair*, July 2008, p. 96.

69   the Mother of All Demos: Nielson, *A Heritage of Innovation: SRI's First Half Century*, p. II-20.

69   "I don't know why we call it a mouse": "The Devices," *The "Mother of All Demos"*: December 9, 1968, video clip on the Web site of SRI International, http://www.sri.com/news/storykits/1968video.html (accessed January 15, 2009).

70   "I looked up and everyone was standing, cheering like crazy": Douglas Engelbart, "The Click Heard Round the World," *Wired*, January 2004, online at http://www.wired.com/wired/archive/12.01/mouse.html (accessed March 24, 2009).

70   astounding event; changed his *life*: Nielson, *A Heritage of Innovation: SRI's First Half Century*, p. II-20.

70   "at least four large computers": Mitch Waldrop, "DARPA and the Internet Revolution," *DARPA: 50 Years of Bridging the Gap*, p. 78.

71   "kind of an 'Aha idea'"; "pretty much instantly made a budget change"; and related quotes: Mayo and Newcomb, "How the Web Was Won," p. 96.

72   Roberts coming to ARPA and other details of the ARPANET project: Hafner and Lyon, *Where Wizards Stay Up Late*; and *The Advanced Research Projects Agency 1958–1974*.

73   "get blamed for the whole goddamned Internet": Mayo and Newcomb, "How the Web Was Won," p. 96.

73   "1969 was quite a year": Ibid.

74   TCP developed: Nielson, *A Heritage of Innovation: SRI's First Half Century*, p. III-6.

75   August 27, 1976, at the Alpine Beer Garden: Ibid., pp. III-1–8.

75   "somewhere between a third and a half of the major innovations in computer science": Michael L. Dertouzos, *What Will Be: How the New World of Information Will Change Our Lives*, New York: HarperCollins, 1997, p. 36.

76   antiwar pins: Hafner and Lyon, *Where Wizards Stay Up Late*, p. 113.

76   "If I could have gotten $30 million from the Red Cross" and university protests: *The Advanced Research Projects Agency 1958–1974*, p. VIII-63.

76   Illiac IV moved to NASA at Ames Research Center: Ed Thelen, "Illiac IV," http://ed-thelen.org/comp-hist/vs-illiac-iv.html (accessed March 24, 2009).

76   "This is *ARPA*": *The Advanced Research Projects Agency 1958–1974*, p. VIII-67.

76   "The military has been much more accepting": Mark Burstein of Boston computer consultancy BBN Technologies, e-mail correspondence, April 24, 2009.

77   SRI history: Nielson, *A Heritage of Innovation: SRI's First Half Century*.

77   SRI charter: Ibid., front matter.

78   SRI promotional video: *Celebrating 60 Years of Innovation,* SRI International, Menlo Park, Calif., 2006.

78   my visit in the summer of 2008: I visited SRI and conducted my interviews there on July 31 and August 1, 2008.

81   "Fourier transform": "How Speech Recognition Works," Aadil Mansoor, http://project.uet.itgo.com/speech.htm (accessed March 24, 2009).

84   David Gunning background and quotes: David Gunning, telephone interview, March 17, 2008.

86   "The direction that personal computing is taking is to help make us more versatile in what we can do easily": Nielson, *A Heritage of Innovation: SRI's First Half Century*, p. II-27.

04: THE ROBOT WILL SEE YOU NOW

95   The patient: I visited the SRI Trauma Pod lab and interviewed Thomas Low on July 31, 2008. More details of the Trauma Pod project and SRI's involvement in it come from Nielson, *A Heritage of Innovation: SRI's First Half Century*, pp. V-1–6, and a telephone interview with Richard Satava, former DARPA program manager, on August 20, 2008.

96   Thomas Low personal details: "Meet Tom Low," Web site of SRI International, http://www.sri.com/about/people/low.html (accessed March 25, 2009).

99   "just flick the switch and do it on the actual patient": Richard Satava, tele-
      phone interview, August 20, 2008.

100   $250,000 for his project: Nielson, *A Heritage of Innovation: SRI's First Half
      Century*, p. V-2.

100   "world's first Telepresence Artist" and other details of SRI's early work on
      medical robotics: Nielson, *A Heritage of Innovation: SRI's First Half Century*, p.
      V-2.

101   Satava called the program MedFast: "21st Century Medicine," *Scientific
      American Frontiers*, air date April 3, 1996, online at http://www.pbs.org/saf/
      previous2.htm (accessed March 25, 2009).

101   ATL Ultrasound spinoff company: "About the Company/History of
      SonoSite," Web site of SonoSite, http://www.sonosite.com/about/history/
      (accessed March 25, 2009).

102   Trauma Pod based on SRI surgical robots: Nielson, *A Heritage of Innovation:
      SRI's First Half Century*, p. V-1.

102   LSTAT: "Products: LSTAT G5," Web site of Integrated Medical Systems,
      http://www.lstat.com/Content/Products.asp (accessed March 25, 2009).

102   Apollo Project and explanation of digital X-ray technology: Bob Kolasky, "A
      Clear View of Innovation: Mammography Goes Digital," online at http://
      www.hks.harvard.edu/sed/docs/k4dev/kolasky_GEcase_2002.pdf (accessed
      August 3, 2008).

102   GE won the runoff: Ibid.

102   replacing conventional X-ray film: Philip J. Hilts, "Digital X-Ray Systems to
      Replace Old Films with Electronic Images," *New York Times*, September 30,
      1997.

102   "the world's first digital mammography system": From a GE Medical Systems
      press release dated April 4, 2000, quoted by Kolasky in "A Clear View of In-
      novation: Mammography Goes Digital."

102   shaving a full one to two years: Ibid.

103   striking similarity to the autodoc: Niven, *World of Ptavvs*, p. 150.

104   Fred Moll: Steve Kichen, "Medical Renaissance," *Forbes*, July 27, 2005,
      online at http://www.forbes.com/2005/07/26/surgery-science-medicine-
      cz_sk_0727surgery.html (accessed March 25, 2009).

104   "new thing called minimally invasive surgery": Satava recalls that it was he
      who reported on advances in laparoscopic surgery to the SRI team, whose
      members then discussed ideas for using the Green Surgical System for it—as
      early as 1987 or 1988. Richard Satava, e-mail correspondence, April 30, 2009.

105   "*repair* of laparoscopic injury": Andrew Kagan, telephone interview, August
      20, 2008.

105   *New England Journal of Medicine* article: Leigh Newmayer, et al., "Open Mesh

versus Laparoscopic Mesh Repair of Inguinal Hernia," *New England Journal of Medicine*, April 29, 2004, p. 1819.

106 *Journal of Clinical Oncology* article: Jim C. Hu, "Utilization and Outcomes of Minimally Invasive Radical Prostatectomy," *Journal of Clinical Oncology*, May 10, 2008, p. 2278.

106 *New York Times* article: Nicholas Bakalar, "Mixed Outcomes in Laparoscopy for Prostates," *New York Times*, May 27, 2008, p. F7.

107 new company called Intuitive Surgical, Inc.: "SRI International's M7 Robot to Perform First-Ever Surgical Demonstration in Zero-Gravity Flight," SRI press release, September 21, 2007, online at http://www.sri.com/news/releases/092107.html (accessed March 25, 2009).

107 $1 million da Vinci robot: "da Vinci System," *Robotic Surgery*, material for class project prepared for Organ Replacement course at Brown University, spring 2005 semester, online at http://biomed.brown.edu/Courses/BI108/BI108_2005_Groups/04/davinci.html (accessed March 25, 2009).

107 da Vinci used in Saudi Arabia and other countries: "Frequently Asked Questions," Web site of Intuitive Surgical, http://www.intuitivesurgical.com/products/faq/index.aspx (accessed March 25, 2009).

111 SNS name: Gregory Myers, Program Director, Engineering and Systems Division, SRI, interview at SRI International, July 31, 2008.

117 cricothyroidotomy: "Cricothyroidotomy," *Encyclopedia of Surgery: A Guide for Patients and Caregivers*, http://www.surgeryencyclopedia.com/Ce-Fi/Cricothyroidotomy.html (accessed March 25, 2009).

119 December 2007 C-9 jet flight: "SRI International's M7 Robot to Perform First-Ever Surgical Demonstration in Zero-Gravity Flight."

122 Mars averages four light-minutes from Earth: Gene Smith, "Gene Smith's Astronomy Tutorial: A Trip Through the Universe at the Speed of Light," Center for Astrophysics and Space Sciences, University of California, San Diego, http://cass.ucsd.edu/public/tutorial/Intro.html (accessed March 25, 2009).

123 University of Cincinnati surgeon Timothy Broderick: Timothy Broderick, telephone interview, August 22, 2008.

124 "the robot will do it for him without error": Richard Satava, telephone interview, August 20, 2008.

CHAPTER 5: BACKSEAT DRIVERS

128 I once asked her: I spoke with Jan Walker at DARPA headquarters on January 4, 2008.

128 Urban Challenge: I made my observations, collected most of my information,

and conducted most of my related interviews during the National Qualifying Event, the race itself, and postrace events in Victorville, California, October 25 to November 4, 2007.

128   With an assignment in hand: I blogged the Urban Challenge for the *Danger Room* on Wired.com, http://blog.wired.com/defense/urban_challenge/index.html (accessed March 25, 2009).

130   Urban Challenge prizes: "DARPA Finalizes 2007 Urban Challenge Cash Prize Levels," DARPA press release, December 8, 2006, online at http://www.darpa.mil/grandchallenge/docs/prize.pdf (accessed March 25, 2009).

132   DARPA autonomous auto races, Norm Whitaker's background, and related quotes: Norm Whitaker, telephone interview, March 19, 2008.

133   Kurjanowicz: "The Great Robot Race," *Nova*, PBS, television, airdate March 28, 2006, online at http://www.pbs.org/wgbh/nova/transcripts/3308_darpa.html (accessed March 25, 2009).

134   Liaison officers: Jan Walker, e-mail correspondence, May 4, 2009.

137   MTVR specs: MTVR brochure, online at http://www.oshkoshdefense.com/pdf/oshkosh_MTVR_brochure.pdf (accessed May 4, 2009).

138   Thrun background: "Sebastian Thrun, Ph.D.," résumé online at http://robots.stanford.edu/cv.pdf (accessed March 25, 2009).

140   Team Gray's Ford Escape hybrid SUV: "Team Gray Technical Paper," DARPA Web site, http://www.darpa.mil/grandchallenge05/TechPapers/GreyTeam.pdf (accessed March 25, 2009).

147   Boss's computer servers and software: Douglas Mechaber, "BOSS—The SUV That Drives Itself," *Tom's Guide*, http://www.tomsguide.com/us/ces-boss-suv,review-1048.html (accessed March 25, 2009).

148   Red Whittaker is a former marine: "The Great Robot Race," *Nova*, PBS.

148   Other details of Red Whittaker's background and related quotes: Red Whittaker, telephone interview, June 20, 2008.

149   Red Whittaker's robots: "Robots at the FRC," Web site of the Field Robotics Center at the Robotics Institute at Carnegie Mellon University, http://www.frc.ri.cmu.edu/robots/index.php (accessed October 21, 2008).

149   Thick gray smoke: "The Great Robot Race," *Nova*, PBS.

151   Golem Group car specs and mishap: Richard Mason, Golem Group team leader, telephone interview, October 16, 2007; "Video: Golem Group Talks About Their Bot's Crash," *TG Daily* video, http://www.tgdaily.com/content/view/34622/113/ (accessed March 25, 2009).

152   "It really is a community": Red Whittaker, telephone interview, June 20, 2008.

159   MIT's maroon Land Rover: Matt Atone et al., "Team MIT Urban Challenge Technical Report," online at http://dspace.mit.edu/bitstream/handle

/1721.1/39822/MIT-CSAIL-TR-2007-058.pdf?sequence=1 (accessed March 25, 2009).

160   "canned routines": Norm Whitaker, telephone interview, March 19, 2008.

160   Norm Whitaker postrace analysis: Norm Whitaker, telephone interview, March 19, 2008.

161   Average speeds of the race winners and accompanying quotes from Tether: Tony Tether, postrace press conference, November 4, 2007.

163   GM keynote: Yi-Wyn Yen, "GM Chief Introduces Greener, Safer Cars at CES Debut," Web site of *Fortune* magazine, http://techland.blogs.fortune.cnn .com/2008/01/09/gm-chief-introduces-greener-safer-cars-at-ces-debut/ (accessed March 25, 2009); Ray Wert, "Rick Wagoner Rocks Down to the Electric Avenue of CES," Jalopnik, http://jalopnik.com/342460/rick-wagoner-rocks-down-to-the-electric-avenue-of-ces (accessed March 25, 2009).

165   Dean Collins quotes: Dean Collins, e-mail correspondence, December 6, 2007.

166   Networked autonomous cars: Rick Wagoner, quoted by Ray Wert, "Rick Wagoner Rocks Down to the Electric Avenue of CES."

06: CRAZY-ASS THINGS

168   2007 DARPATech: DARPATech 2007 took place at the Anaheim Marriott Hotel in Anaheim, Calif., August 7–9, 2007.

168   David Lucia on DARPA as Disneyland: David Lucia, program manager, DARPA Tactical Technology Office recruitment video, online at http://www .darpa.mil/hrd/videos.htm# (accessed February 7, 2009).

168   DARPA program managers serve four- to six-year terms: Jan Walker, e-mail correspondence, May 4, 2009.

169   Noah Shachtman recruited by a DARPA program manager: Noah Shachtman, "DARPA Brain Drain Costs Agency $32 Million," *Danger Room*, Wired. com, http://blog.wired.com/defense/2008/06/the-us-military.html, June 8, 2008; Noah Shachtman, telephone interview, June 24, 2008.

169   sidebar discussions: "About Sidebar Discussions," prepared for DARPATech 2007, http://www.darpatechsidebars.com/files/Sidebar_Guide.pdf (accessed February 5, 2009).

170   USC speech researchers: The head of the project at USC was Shrikanth S. Narayanan, and the technical lead was Panayiotis Georgiou, both of the Viterbi School of Engineering.

171   computer chips in insects: Sean Meade, "Return of the Cyborg Moths," Web site of *Aviation Week* magazine, http://www.aviationweek.com/aw/ blogs/defense/index.jsp?plckController=Blog&plckScript=blogScript &plckElementId=blogDest&plckBlogPage=BlogViewPost&plckPostId=

Blog%3A27ec4a53-dcc8-42d0-bd3a-01329aef79a7Post%3Aa1585642-ac31-4fb3-be20-e3ba53cf86fb; Tsuneyuki Miyake, "US University Shows Radio-controlled Live Beetle," *Tech-On!*, http://techon.nikkeibp.co.jp/english/NEWS_EN/20090128/164717/; Noah Shachtman, "Pentagon's Cyborg Beetle Spies Take Off," *Danger Room*, Wired.com, http://blog.wired.com/defense/2009/01/pentagons-cybor.html #more (all accessed March 25, 2009).

171   Amit Lal presentation: Amit Lal, "Micro and Nano Electro-Mechanical Systems: Technology Engineering Metamorphosis," online at http://www.darpa.mil/DARPAtech2007/proceedings/dt07-mto-lal-micronano.pdf (accessed March 25, 2009).

172   "energy has been *the* limiting factor in all military operations": Douglas Kirkpatrick, "Energy as a Tactical Asset," DARPATech 2007 presentation, online at http://www.darpa.mil/DARPAtech2007/proceedings/dt07-sto-w-kirkpatrick-energy.pdf (accessed March 25, 2009).

172   Benjamin Mann: Benjamin Mann, "DSO Mathematics: The Heart and Soul of the Far Side," DARPATech 2007 presentation, online at http://www.darpa.mil/DARPAtech2007/proceedings/dt07-dso-mann-mathematics.pdf (accessed March 25, 2009).

173   titanium, $35.00 per pound to $3.50 per pound: Brett Giroir, director, DARPA Defense Sciences Office, "IDEAS Begin Here," DARPATech 2007 presentation, online at http://www.darpa.mil/DARPAtech2007/proceedings/dt07-dso-giroir-ideas.pdf (accessed March 25, 2009).

173   landing at the National Geographic Channel: The program aired as part of the network's *Inside* series, as *America's Secret Weapons*, on December 4, 2008.

175   Tony Tether background and quotes: Tony Tether, interview at DARPA headquarters, Arlington, Va., January 4, 2008, and telephone interview, June 25, 2008.

175   Magazine with my cover story on a DARPA project: Michael Belfiore, "The Hypersonic Age Is Near, *Popular Science*, January 2008, p. 36, online at http://www.popsci.com/military-aviation-space/article/2007-12/hypersonic-age-near (accessed March 25, 2009).

176   "If you're going to create surprise": Tony Tether, interview at DARPA headquarters, May 6, 2008.

176   Development of stealth technology: "Stealth Technology," *Modern Marvels*, History Channel video, 1997.

178   F-22 Raptor made them obsolete: Peter Pae, "Stealth Fighters Fly off the Radar," *Los Angeles Times*, April 23, 2008, online at http://articles.latimes.com/2008/apr/23/business/fi-stealth23 (accessed March 25, 2009).

178   DARPA researcher from the Unied Kingdom: The researcher's name is Phil Hunt.

179   "intellectually aggressive": Douglas Kirkpatrick, telephone interview, April 4, 2008.

180   25 percent changeover rate every year: Jan Walker, e-mail correspondence, May 4, 2009.

181   Clarke paper: Arthur C. Clarke, "Extra-Terrestrial Relays: Can Rocket Stations Give World-Wide Radio Coverage?" *Wireless World*, October 1945, p. 305, online at http://www.clarkefoundation.org/docs/ClarkeWireles WorldArticle.pdf (accessed March 25, 2009).

182   "Someone who says something can't be done and can't prove it is someone to be avoided by DARPA": Joe Mangano, quoted by Barry Rosenberg, "DARPA Paves the Way for U.S. Efforts in Ballistic Missile Defense," *DARPA: 50 Years of Bridging the Gap*, p. 68.

182   ESP research: H. E. Puthoff, "CIA-Initiated Remote Viewing at Stanford Research Institute," online at http://www.biomindsuperpowers.com/Pages/ CIA-InitiatedRV.html (accessed March 25, 2009); Francine du Plessix Gray, "Parapsychology and Beyond," *New York Times Magazine*, August 11, 1974, p. 219.

182   "tremendous if you could do it": Charles Piller, "Column One; Army of Extreme Thinkers; The Brilliant Successes of DARPA, the Defense Department's Advanced Research Agency, Are Matched Only by Its Long List of Bizarre Failures," *Los Angeles Times*, August 14, 2003, p. A1.

183   ballistic missile defense: Rosenberg, "DARPA Paves the Way for U.S. Efforts in Ballistic Missile Defense," pp. 64–73.

184   "think a lot before you draw the sword": Ibid., p. 72.

184   Halide excimer lasers: Ibid., p. 67.

184   Arecibo radio telescope: Ibid., p. 64.

185   "unclear to the whole Department of Defense": Tony Tether, telephone interview, May 6, 2008.

186   "I was a Fuller Brush man": Ibid.

187   Manpacks: "Manpack Global Positioning System (GPS) Receiver," Web site of the National Museum of American History, http://americanhistory.si.edu/ collections/object.cfm?key=35&objkey=220; "Rockwell Manpack Global Positioning System (GPS) Receiver," Navigation Museum, Institute of Navigation, http://www.ion.org/museum/item_view.cfm?cid=7&scid=9&iid=10 (both accessed March 25, 2009).

187   STO program manager Sherman Karp and GPS: Alexandrow, "The Story of GPS," p. 54.

187   Details of handheld GPS genesis: Tony Tether, interview at DARPA headquarters, Arlington, Va., May 6, 2008.

187   Rockwell International's GPS assignment and Ron Coffin quotes: Jill

Brimeyer, "You've Come a Long Way, GPS," Web site of Jill Brimeyer, http://jillbrimeyer.com/GPSstory.html (accessed March 25, 2009).

188   "It led us to precision missiles": Tony Tether, interview at DARPA headquarters, Arlington, Va., June 25, 2008.

189   "financial bath": Tony Tether, telephone interview, May 6, 2008.

190   Rumsfeld Commission report: Report of the Commission to Assess United States National Security Space Management and Organization, online at http://www.dod.mil/pubs/space20010111.html (accessed May 7, 2009).

193   TIA and Poindexter: John Markoff, "Chief Takes over at Agency to Thwart Attacks on U.S.," *New York Times*, February 13, 2002, online at http://query.nytimes.com/gst/fullpage.html?res=9D00E0D61F3CF930A25751C0A9649C8B63 (accessed March 25, 2009).

193   Poindexter's immunity and appeal: Clyde Haberman, "Arthur L. Liman, a Masterly Lawyer, Dies at 64," *New York Times*, July 18, 1997, online at http://query.nytimes.com/gst/fullpage.html?res=9C0DEED71E38F93BA25754C0A961958260&sec=&spon=&pagewanted=all (accessed March 25, 2009).

194   The beginning of the end: William Safire, "You Are a Suspect," *New York Times*, November 14, 2002, online at http://query.nytimes.com/gst/fullpage.html?res=9F0CE6D71630F937A25752C1A9649C8B63 (accessed March 25, 2009).

195   "I can't think of a good countermeasure": Audrey Hudson, "Homeland Bill 'a Supersnoop's Dream,'" *Washington Times*, November 15, 2002, p. A1.

195   TIA report: Report to Congress Regarding the Terrorism Information Awareness Program, Executive Summary, online at http://www.information-retrieval.info/docs/tia-exec-summ_20may2003.pdf (accessed March 25, 2009).

197   Leonard Kleinrock: John Markoff, "University Scientists Concerned by Cuts in Computer Projects," *New York Times*, April 2, 2005, p. C1, online at http://www.nytimes.com/2005/04/02/technology/02darpa.html?scp=1&sq=kleinrock%20darpa&st=cse (accessed March 25, 2009).

197   "open-ended research is categorized as 6.1": Tony Tether, e-mail correspondence, April 30, 2009.

198   "if I get fired, he gets fired": Douglas Kirkpatrick, telephone interview, April 4, 2008.

198   fiftieth-anniversary celebration: I attended the celebration on April 10, 2008, at the Washington Hilton, in Washington, D.C., and filed a report for the Web site of *Popular Science*, at http://www.popsci.com/military-aviation-space/article/2008-04/darpa-turns-50 (accessed March 25, 2009).

199   "Dr. Tether will be here after January 20": Jan Walker, e-mail correspondence, January 6, 2009.

199   "DARPA is envied": Tony Tether, e-mail correspondence, April 30, 2009.

07: THE FINAL FRONTIER

201 late summer day in Hampton, Virginia: On September 3, 2008, I visited NASA's Langley Research Center, watched a presentation by aerospace engineers, and conducted interviews.

201 F-22 facts: "F-22 Raptor," Air Force Link, Web site of the U.S. Air Force, http://www.af.mil/factsheets/factsheet.asp?fsID=199 (accessed March 26, 2009).

201 NASA Langley history: "Langley History," Web site of the National Aeronautics and Space Administration, http://www.nasa.gov/centers/langley/about/history.html (accessed March 26, 2009).

202 past five times the speed of sound: Michael Belfiore, "The Hypersonic Age Is Near, *Popular Science*, January, 2008, p. 36, online at http://www.popsci.com/military-aviation-space/article/2007-12/hypersonic-age-near (accessed March 25, 2009).

202 theoretical top speed of Mach 15: Thomas A. Jackson, "Power for a Space Plane," *Scientific American*, August 2006, p. 56.

204 "higher specific impulse" and Steven Walker quotes and background: Steven Walker, telephone interview, March 17, 2008.

204 Reagan State of the Union address: Ronald Reagan, "Address Before a Joint Session of the Congress Reporting on the State of the Union," www.Reagan2020.us, at http://reagan2020.us/speeches/state_of_the_union_1986.asp (accessed March 26, 2008).

204 NASP: John Pike, "X-30 National Aerospace Plane (NASP)," *Mystery Aircraft— Our Analysis*, Web site of the Federation of American Scientists, at http://www.fas.org/irp/mystery/nasp.htm (accessed March 26, 2009).

205 NASP canceled: Timothy Gaffney, "X-30 Superplane Becomes 8-Year Flight of Fancy," *Dayton Daily News*, February 15, 1994, p. 1A.

205 "We didn't stop our scramjet research": Charlie Brink, telephone interview, September 27, 2007.

206 "we need faster missile platforms": Steven Walker, telephone interview, March 17, 2008.

207 "transition of the technology out of the laboratories": David Van Wie, telephone interview, October 11, 2007.

208 HyShot: "HyShot," Web site of the Centre for Hypersonics at the University of Queensland, Australia, at http://www.uq.edu.au/hypersonics/?page=19501 (accessed March 26, 2009).

208 X-43A: X-43X project page, Web site of the National Aeronautics and Space Administration, at http://www.nasa.gov/missions/research/x43-main.html (accessed March 26, 2009).

211–212 SpaceX and Musk quote: Belfiore, *Rocketeers*.

213 AirLaunch LLC: Ibid.; Web site of AirLaunch LLC, at http://airlaunchllc.com/ (accessed March 26, 2009).

214   Minotaur: "Minotaur Space Launch Vehicles," Web site of Orbital Sciences
       Corporation, at http://www.orbital.com/SpaceLaunch/Minotaur/ (accessed
       March 26, 2009).

215   "have flown hypersonic demonstrators in the past": Steven Walker, interview
       at DARPA headquarters, Arlington, Va., June 25, 2008.

216   HyFly: "Hypersonic Flight Demonstration (HyFly)," DARPA Web site, at
       http://www.darpa.mil/TTO/Programs/hyfly.htm; Graham Warwick, "Boe-
       ing's HyFly Hypersonic Missile Fails in Bid for Mach 6," *Flight Global*, Novem-
       ber 2, 2008, at http://www.flightglobal.com/articles/2008/02/11/221478/
       boeings-hyfly-hypersonic-missile-fails-in-bid-for-mach-6.html; "HyFly: Hy-
       personic Flight Demonstration," Web site of the Office of Naval Research,
       at http://www.onr.navy.mil/media/extra/fact_sheets/hyfly.pdf (all accessed
       on March 26, 2009).

217   "We tested it at Mach 4.6, 5.0, and 6.5," Pratt and Whitney Rocketdyne's
       tests: Curtis Berger and Mike McKeon, PWR managers, telephone interview,
       October 3, 2007.

219   Wind tunnel details: From my visit to the facility on September 3, 2008,
       and Lawrence D. Huebner, et al., "Calibration of the Langley 8-Foot High
       Temperature Tunnel for Hypersonic Airbreathing Propulsion Testing,"
       American Institute of Aeronautics and Astronautics, online at http://www
       .tpub.com/content/nasa1996/NASA-aiaa-96-2197/NASA-aiaa-
       96-21970002.htm; Pete Jacobs, "8-Foot High Temperature Tunnel," NASA Web
       site, http://windtunnels.larc.nasa.gov/facilities_updated/hypersonic/8ft.
       htm; Jeffrey S. Hodge and Stephen F. Harvin, "The NASA Langley Research
       Center 8-ft High Temperature Tunnel," Advanced Hypersonic Test Facilities
       (Progress in Astronautics and Aeronautics, Vol. 197), American Institute of
       Aeronautics and Astronautics, p. 405, online at http://books.google.com/
       books?id=_D1BJ_wlAv4C&pg=RA1-PA405&lpg=RA1-PA405&dq=nasa+la
       ngley+htt&source=web&ots=0xXsU5rnOa&sig=tcsYYTUA1dSy-xNBQW
       XQSbg34xQ&hl=en&sa=X&oi=book_result&resnum=7&ct=result#PRA1-
       PA408,M1; "Wind Tunnels of NASA," http://www.hq.nasa.gov/office/pao/
       History/SP-440/ch6-16.htm (all accessed March 26, 2009).

221   Countdown and test-firing details: Video of a test on March 23, 2007, pro-
       vided by Pratt and Whitney Rocketdyne.

223   "People like to talk about 'DARPA-hard'": Steven Walker, interview at
       DARPA headquarters, Arlington, Va., June 25, 2008.

224   Blackswift budget cut from $120 million to $10 million, and related quote
       from Steven Walker: Stephen Trimble, "VIDEOS: DARPA Cancels Blackswift
       Hypersonic Test Bed," October 13, 2008, *Flight Global*, http://www.flight-
       global.com/articles/2008/10/13/317382/videos-darpa-cancels-blackswift-
       hypersonic-test-bed.html (accessed March 26, 2009).

224 "a lot more going on than just the few papers": Craig Covault, telephone interview, September 28, 2007.

225 LAPCAT: Michael Belfiore, "Green Skies at Mach 5," *Popular Science*, February 2008, p. 48, online at http://www.popsci.com/military-aviation-space/article/2008-01/green-skies-mach-5 (accessed March 26, 2009); Richard Varvill, technical director and chief designer of Reaction Engines, telephone interview, November 12, 2007; Remy Demos, LAPCAT project officer at the European Commission, telephone interview, November 14, 2007.

226 Concorde introduced in 1969: "1969: Concorde flies for the first time," BBC Web site, http://news.bbc.co.uk/onthisday/hi/dates/stories/march/2/newsid_2514000/2514535.stm (accessed March 26, 2009).

227 "development timescale that appears to be on the order of decades": David Van Wie, telephone interview, October 11, 2007.

228 "taking the technology excuse off the table": Steven Walker, interview at DARPA headquarters, Arlington, Va., June 25, 2008.

CHAPTER 8: POWER TO THE PEOPLE

229 "tactical energy independence" and related quotes and details: Douglas Kirkpatrick, "Energy as a Tactical Asset," DARPATech 2007 presentation, online at http://www.darpa.mil/DARPAtech2007/proceedings/dt07-sto-w-kirkpatrick-energy.pdf (accessed March 25, 2009).

229 "We are a strategic agency," related details, quotes, and Kirkpatrick's background: Douglas Kirkpatrick, telephone interview, April 4, 2008.

232 HEDLight: "HEDLight," DARPA Web site, http://www.darpa.mil/sto/maritime/hedlight.html (accessed March 26, 2009).

233 Very High Efficiency Solar Cells program: Richard Stevenson, "Photovoltaics Take a Load off Soldiers," *Compound Semiconductor*, October 2006, p. 28; Eli Kintisch, "Light-Splitting Trick Squeezes More Electricity out of Sun's Rays," *Science*, August 3, 2007, p. 583; Allen Barnett, Douglas Kirkpatrick, and Christiana Honsberg, "New US Ultra High Efficiency R&D Programme," paper presented at the Twenty-first European Photovoltaic Solar Energy Conference, Dresden, Germany, September 4, 2006.

233 Advantages of nanoscale structures in solar cells: Nancy Stauffer, "Nanoscale Layers Promise to Boost Solar Cell Efficiency," Web site of MIT Energy Initiative, http://web.mit.edu/mitei/research/spotlights/nano-layers.html (accessed March 26, 2009).

235 20 percent efficiency of solar cells on the market: Bennett Daviss, "Our Solar Future," *New Scientist*, December 8, 2007, p. 32.

236 Fifteen universities and private labarotories: Neil Thomas, "UD to Lead $53

Million Solar Cell Initiative," *UDaily*, November 2, 2005, online at http://www.udel.edu/PR/UDaily/2006/nov/solar110205.html (accessed March 26, 2009).

236   Allen Barnett and Christiana Honsberg quotes, related details, and background: Allen Barnett and Christiana Honsberg, interview at the University of Delaware, May 22, 2008.

238   MBE machine: I visited the University of Delaware to interview Barnett and Honsberg and gather details of the solar program, including the MBE machine, on May 22, 2008; Chris Honsberg, "New MBE Machine for Ultra-High Efficiency Solar Cells," online at http://www.ece.udel.edu/~honsberg/MBE.html (accessed March 26, 2009).

240   Dichroic mirrors were Kirkpatrick's idea: Allen Barnett, telephone interview, March 24, 2008.

240   How the solar cells work: Kintisch, "Light-Splitting Trick Squeezes More Electricity Out of Sun's Rays," *Science*, August 3, 2007, p. 583.

241   Ready for production by 2011: Ibid.

241   "This is just a bunch of people making a bunch of hooey" and related quotes: Douglas Kirkpatrick, telephone interview, April 4, 2008.

241   Rudolph Diesel's experiments with biofuels: Gerhard Knothe, "Historical Perspectives on Vegetable Oil-Based Diesel Fuels," *Inform*, vol. 12, November 2001, pp. 1103–4, online at http://www.biodiesel.org/resources/reportsdatabase/reports/gen/20011101_gen-346.pdf (accessed October 31, 2008).

242   Biodiesels contribute little to atmospheric carbon dioxide: John Sheehan, et al., "Life Cycle Inventory of Biodiesel and Petroleum Diesel," study sponsored by the U.S. Department of Agriculture and the U.S. Department of Energy, May 1998, p. 18, online at http://www.nrel.gov/docs/legosti/fy98/24089.pdf (accessed March 26, 2009). The report states "the overall life cycle emissions of CO_2 from B100 [100 percent biodiesel] are 78.45% lower than those of petroleum diesel. The reduction is a direct result of carbon recycling in soybean plants."

243   Drawbacks of conventional biodiesel: Douglas Kirkpatrick, telephone interview, April 4, 2008, and Vern Hoffman, et al., "Biodiesel Use in Engines," North Dakota State University Extension Service publication, January 2006, online at http://www.ag.ndsu.edu/pubs/ageng/machine/ae1305w.htm (accessed March 26, 2009). The paper states, "Commercially available (soybean) biodiesel (B100) will gel at about 30 F."

243   BioFuels criteria: "BioFuels: BAA06-43," DARPA solicitation, online at http://www.darpa.mil/sto/solicitations/BioFuels/ (accessed March 26, 2009).

244   EERC's biofuel tests: Derek Walters, Communications Manager, EERC, e-mail correspondence, April 3, 2009.

244   EERC's head start in biofuels: Ibid.

245   April 2008: I visited the Energy and Environmental Research Center to conduct interviews and see ongoing work on April 14, 2008.

246   "My definition of peer review" and related quotes on the EERC's organizational structure: Gerald Groenewold, telephone interview, March 11, 2008.

246   Groenewold launched EERC in 1974, other background, related quotes, and details of current EERC projects: Gerald Groenewold, interview at EERC, April 14, 2008.

250–51 "We approached it more from a refining process": Tom Erickson, interview at EERC, April 14, 2008.

251   EERC process and other details and quotes: Douglas Kirkpatrick, telephone interview, April 4, 2008.

252   "We're going to have to move away from using food to produce energy": Gerald Groenewold, interview at EERC, Grand Forks, North Dakota, April 14, 2008.

254   "Sometimes the sun won't shine": Kirkpatrick. "Energy as a Tactical Asset," DARPATech 2007 presentation.

257   ARPA-E: Eli Kintisch, "$400 Million for Off-the-Wall Energy Ideas," *ScienceInsider*, February 12, 2009, online at http://blogs.sciencemag.org/scienceinsider/2009/02/400-million-for.html (accessed March 26, 2009).

SELECTED BIBLIOGRAPHY

Applied Physics Laboratory. *The First Forty Years: A Pictorial Account of the Johns Hopkins Applied Physics Laboratory Since Its Founding in 1942* (Baltimore, Md.: Schneidereith and Sons, 1983). Online at http://www.jhuapl.edu/aboutapl/heritage/first-forty/ (accessed March 24, 2009).

Belfiore, Michael. "Green Skies at Mach 5." *Popular Science*, February 2008, p. 48. Online at http://www.popsci.com/military-aviation-space/article/2008-01/green-skies-mach-5 (accessed March 26, 2009).

——. "The Hypersonic Age Is Near." *Popular Science*, January 2008, p. 36. Online at http://www.popsci.com/military-aviation-space/article/2007-12/hypersonic-age-near (accessed March 25, 2009).

——. *Rocketeers: How a Visionary Band of Business Leaders, Engineers and Pilots Is Boldly Privatizing Space* (New York: Smithsonian Books/HarperCollins, 2007).

Boykin, Sam. "With Open-Source Arms." *Scientific American*, October 2008, p. 90.

Brzezinski, Matthew. *Red Moon Rising: Sputnik and the Hidden Rivalries That Ignited the Space Race* (New York: Times Books/Henry Holt and Company, 2007).

Clarke, Arthur C. "Extra-Terrestrial Relays: Can Rocket Stations Give World-Wide Radio Coverage?" *Wireless World*, October 1945, p. 305. Online at http://www.clarkefoundation.org/docs/ClarkeWirelessWorldArticle.pdf (accessed march 25, 2009).

DARPA: 50 Years of Bridging the Gap (Tampa, Fla.: U.S. Department of Defense and Faircount Media Group, 2008).

Eisenhower, Dwight D. *The White House Years: Waging Peace, 1956–1961* (Garden City, N.Y.: Doubleday and Company, 1965).

Engelbart, Douglas. "The Click Heard Round the World." *Wired*, January 2004. Online at http://www.wired.com/wired/archive/12.01/mouse.html (accessed March 24, 2009).

——. *The "Mother of All Demos"*: December 9, 1968. (SRI International video). Online at http://www.sri.com/news/storykits/1968video.html (accessed January 15, 2009).

Giroir, Brett. "IDEAS Begin Here." DARPATech 2007 presentation. Online at http://www.darpa.mil/DARPAtech2007/proceedings/dt07-dso-giroir-ideas.pdf (accessed March 25, 2009).

Hafner, Katie, and Matthew Lyon. *Where Wizards Stay Up Late* (New York: Simon and Schuster, 1996).

Harris, Gordon L. *Selling Uncle Sam: A Firsthand Look at the Public Relations Behind the U.S. Space Program* (Hicksville, N.Y.: Exposition Press, 1976).

Hilts, Philip J. "Digital X-Ray Systems to Replace Old Films with Electronic Images." *New York Times*, September 30, 1997, p. F9.

History Channel. *Modern Marvels: Stealth Technology* (1996).

Kichen, Steve. "Medical Renaissance." *Forbes*, July 27, 2005. Online at http://www.forbes.com/2005/07/26/surgery-science-medicine-cz_sk_0727surgery.html (accessed March 25, 2009).

Killian, Jr., James R. *Sputnik, Scientists and Eisenhower* (Cambridge, Mass.: MIT Press, 1977).

Kintisch, Eli. "$400 Million for Off-the-Wall Energy Ideas." *ScienceInsider*, February 12, 2009, http://blogs.sciencemag.org/scienceinsider/2009/02/400-million-for.html (accessed March 26, 2009).

——. "Light-Splitting Trick Squeezes More Electricity out of Sun's Rays." *Science*, August 3, 2007, p. 583.

Kirkpatrick, Douglas. "Energy as a Tactical Asset." DARPATech 2007 presentation. Online at http://www.darpa.mil/DARPAtech2007/proceedings/dt07-sto-w-kirkpatrick-energy.pdf (accessed March 25, 2009).

Kolasky, Bob. "A Clear View of Innovation: Mammography Goes Digital." Online at http://www.hks.harvard.edu/sed/docs/k4dev/kolasky_GEcase_2002.pdf (accessed August 3, 2008).

Lal, Amit. "Micro and Nano Electro-Mechanical Systems: Technology Engineering Metamorphosis." DARPATech 2007 presentation. Online at http://www.darpa.mil/DARPAtech2007/proceedings/dt07-mto-lal-micronano.pdf (accessed March 25, 2009).

Licklider, J. C. R. "Man-Computer Symbiosis." *IRE Transactions on Human Factors in Electronics*, Volume HFE-1, March 1960, pp. 4–11. Online at http://groups.csail.mit.edu/medg/people/psz/Licklider.html (accessed March 24, 2009).

Mann, Benjamin. "DSO Mathematics: The Heart and Soul of the Far Side." DARPATech 2007 presentation. Online at http://www.darpa.mil/DARPAtech2007/proceedings/dt07-dso-mann-mathematics.pdf (accessed March 25, 2009).

Markoff, John. "University Scientists Concerned by Cuts in Computer Projects." *New York Times*, April 2, 2005, p. C1. Online at http://www.nytimes.com/2005/04/02/technology/02darpa.html?scp=1&sq=kleinrock%20darpa&st=cse (accessed March 25, 2009).

Mayo, Keenan, and Peter Newcomb. "How the Web Was Won." *Vanity Fair*, July 2008, p. 96.

Medaris, J. B.: *Countdown for Decision* (New York: G. P. Putnam Sons, 1960).

Murphy, Charles J. V. "The White House Since Sputnik." *Fortune*, January 1958, p. 98.

Neufeld, Michael J. *Von Braun: Dreamer of Space, Engineer of War* (New York: Alfred A. Knopf, 2007).

Nielson, Donald L. *A Heritage of Innovation: SRI's First Half Century* (Menlo Park, Calif.: SRI International, 2004, 2006).

Niven, Larry. *The World of Ptavvs: Three Books of Known Space* (New York: Ballantine Books, 1966, 1968, 1975, 1990, 1996, renewed 1994, 1996).

"The Organization Man." *Time*, January 13, 1958. Online at http://www.time.com/time/magazine/article/0,9171,862823,00.html (retrieved May 1, 2008).

Piller, Charles. "Column One; Army of Extreme Thinkers; The Brilliant Successes of DARPA, the Defense Department's Advanced Research Agency, Are Matched Only by Its Long List of Bizarre Failures." *Los Angeles Times*, August 14, 2003, p. A1.

Public Broadcasting System. *Nova: The Great Robot Race*, aired March 28, 2006. Online at http://www.pbs.org/wgbh/nova/transcripts/3308_darpa.html (accessed March 25, 2009).

——. *Scientific American Frontiers: 21st Century Medicine*, aired April 3, 1996. Online at http://www.pbs.org/saf/previous2.htm (accessed March 25, 2009).

Raymond, Jack. "Pentagon Forms a Space Agency; G.E. Aid Is Chief." *New York Times*, February 8, 1958, p. 1.

Richard J. Barber Associates. *The Advanced Research Projects Agency 1958–1974*. (Washington, D.C.: Advanced Research Projects Agency, 1975). Online at http://handle.dtic.mil/100.2/ADA154363 (accessed March 24, 2009).

Rumsfeld, Donald, et al. *Report of the Commission to Assess United States National Security Space Management and Organization*, January 11, 2001. Online at http://www.dod.mil/pubs/space20010111.html (accessed May 7, 2009).

Safire, William. "You Are a Suspect." *New York Times*, November 14, 2002. Online at http://query.nytimes.com/gst/fullpage.html?res=9F0CE6D71630F937A25752C1A9649C8B63 (accessed March 25, 2009).

"Transcript of the President's News Conference on Foreign and Domestic Matters." *New York Times*, October 10, 1957, p. 14.

von Braun, Wernher. "The Story Behind the Explorers." *This Week*, April 13, 1958, p. 8.

Weisskopf, Michael. *Blood Brothers*. (New York: Henry Holt and Company, 2006).

"The Young Rocketeers." *Time*, January 13, 1958. Online at http://www.time.com/time/magazine/article/0,9171,862854,00.html (accessed May 1, 2008).

INDEX

ALSO BY
MICHAEL BELFIORE

ROCKETEERS

How a Visionary Band of Business Leaders, Engineers, and Pilots is Boldly Privatizing Space

ISBN 978-0-06-114903-0 (paperback)

The first book to tell the story of the people who are succeeding in making private space travel a reality. In this entertaining narrative, Belfiore takes readers close to people who are not afraid to dream and achieve what seems impossible.

"My walk on the moon was an important beginning, but the privatization of space travel is an essential step toward realizing our cosmic destiny. . . . Michael Belfiore tells the fascinating story of the entrepreneurs who have already made it happen."

—Buzz Aldrin